John Angell

Elements of Animal Physiology, Chiefly Human

John Angell

Elements of Animal Physiology, Chiefly Human

ISBN/EAN: 9783337240004

Printed in Europe, USA, Canada, Australia, Japan

Cover: Foto ©berggeist007 / pixelio.de

More available books at **www.hansebooks.com**

Putnam's Elementary Science Series.

ELEMENTS

OF

ANIMAL PHYSIOLOGY,

CHIEFLY HUMAN.

BY

JOHN ANGELL,

SENIOR SCIENCE MASTER, MANCHESTER GRAMMAR SCHOOL.

ILLUSTRATED WITH 83 FIGURES.

NEW YORK:

G. P. PUTNAM'S SONS,

FOURTH AVENUE AND TWENTY-THIRD STREET.

PREFACE.

ANIMAL PHYSIOLOGY is "essentially a Science of Designs and Final Causes." Probably no subject, if taught with even moderate intelligence, can be rendered so interesting, or made to rouse the dormant energies of the intellect of young people, and of those previously untrained in the method of scientific inquiry, as that of Human Physiology. Certainly the study of no other science opens up to the mind so many wonders, solves so many human mysteries, or brings home to the mind so many proofs of the "reign of Law," and of the wise and beneficent design of this universe, or teaches so many princi-ples which lie at the basis of human health, morals, and well-being, as that of Human Physiology. On these considerations alone,—and many other most powerful ones might be urged, —"Human Physiology" must claim a leading position in any system which professes to educate the people *soundly*.

Very considerable experience as a Teacher of Physiology, with the youthful and adult of both sexes, has proved to the writer the facility with which the *viva voce* teaching of Elemen-tary Physiology may be made "perfectly sound and thorough," when supplemented with the aid of *good diagrams*, a good *text-book*, and the free use of the *black-board*, and of the lungs, heart, kidney, eye, &c., of the sheep, and of such other objects as may be readily obtained from the butcher, for the illustra-tion of the general structure of the corresponding organs of the human body; while the body of the *living* frog or tadpole is amply sufficient to illustrate the general phenomena of the circulation and the properties of the *living* tissues. The teacher should also occasionally dissect a small animal, as a rabbit, recently killed with chloroform, before his class. A Human Skeleton and a cheap Microscope will likewise be found invaluable aids to the earnest teacher.

A general impression prevails, among the candidates pre-senting themselves at the Government Science Examinations, that it is easier to pass with the same actual knowledge in the Second Class, Advanced Stage, than in the First Class, Elementary Stage. The prevalence of this error can only have resulted from the candidates having confined their

attention too exclusively to the subjects *named* in the Syllabus
of the Elementary Stage, whereas it is impossible to obtain a
sound, however elementary, knowledge of these subjects themselves, without their also making themselves familiar, as it
were, with the atmosphere of facts immediately surrounding
them.   To obviate this difficulty, promote greater soundness
of attainment, and especially with the object of cultivating
the higher reasoning powers of the young student in dealing
with the principles of the science, rather than encourage in
him the acquisition of the power of what may be expressed as
mere "descriptive cram," the writer has dealt, as at the opening of the book, and in treating of the blood and the tissues,
far more with the principles and the reasoning of the subject
than is usual in treatises of the same limited and elementary
character.   He has therefore been compelled to extend the
range of this little work beyond the subjects actually named
in the Syllabus of the *Elementary* stage issued by the Depart-
ment of Science and Art.   He has also everywhere sought,—as
by sectionizing the work, by making it as *practical* as possible,
by the mode of keying the diagrams, &c.,—to make it as clear
and thorough as his limits would permit, never hesitating to
repeat, where he felt repetition would be advantageous to the
student.

   The writer begs, in conclusion, to express his indebtedness
to the works of Bennett, Carpenter, Combe, Gray, Lawson,
Marshall, Playfair, Huxley, and others; to the latter especially
for kindly allowing him to copy some of the diagrams in his
text-book on the same subject.   To Professor Huxley all
friends of the cause of scientific education are, in common
with the writer, indebted, not only for his labours as an
Englishman in extending the domains of Biological Science,
but also for his able and successful advocacy of the claims of
"Physiology" as a branch of general popular education.

                                                JOHN ANGELL.

MANCHESTER, *August*, 1873.

CONTENTS.

CHAPTER V. HISTOLOGICAL PRELIMINARIES.

CHAPTER VI. THE BLOOD.

CHAPTER VII. CIRCULATION OF THE BLOOD—THE HEART AND BLOOD-VESSELS.

CHAPTER VIII. RESPIRATION—THE LUNGS.

ANIMAL PHYSIOLOGY.

CHAPTER I.

GENERAL VIEW OF THE SCHEME AND FUNCTIONS OF THE HUMAN BODY.

1. The Living Body compared with a Steam Engine at Work.—The body of a *living* man in many respects closely resembles that of a *steam engine* at *work*, differing from it chiefly in the greater variety, complexity, beauty, and perfection of its individual parts, and its scheme as a whole, just as its contriver and maker, *man*, differs in power and degree from the great Maker of the Universe.

2. Physiology and Theology in Harmony.—At the outset, to prevent all possible confusion, let it be distinctly understood that, though we shall frequently have cause to refer to the nature and the condition of the action of the *human mind*, we in no case intend, as not coming within the proper province of "physiology," to refer to the nature of the *human soul*, or its connections or relations with the human body.

In the opinion of the writer, as far as mere human knowledge and intellect are concerned, the nature of this connection is entirely a mystery: all attempts to investigate this divine, not human, problem have hitherto not only failed to present any light on the wonderful mystery, but have entirely failed even to show us *how*, by any *scientific* or *philosophical* methods with which we are acquainted, we shall be able to acquire such knowledge.

In fact, we at the present time are not only without

such knowledge, in the sense in which the term is used in *science*, but are also utterly helpless as to the *method* by which such knowledge is to be *philosophically* acquired or built up. The writer of this little treatise would even go so far as to express it to be his opinion that the Almighty, in creating man, withheld from him, as he has also withheld from him the *power* of prophecy and other Divine gifts, the faculty of acquiring during this life any true knowledge of the nature of the *human soul.*

3. Some writers, it is true, have endeavoured to support, on the basis of *physiological* argument, the theory of the utter *extinction* and *annihilation* of man by *death.* With this view the writer of this little book has no sympathy whatever. In any case the proposition, being incapable of *scientific proof*, must remain in the region of pure *hypothesis.* To the present writer the fact that the Great Creator of all things has endowed man, as a part of his nature and his *mental being*, with the fundamental faculty or the natural sentiment of religion, in consequence of which mankind, even in the absence of true knowledge, has at all times and during all ages *practised* religion, *true* or *false*—has at all times cultivated a belief in an existence *future* to death—seems (to the writer) distinctly to point to, and make as eminently reasonable, the belief in the truth of the *future existence* or the *immortality* of man, as the *structure* of the *heart* and *great blood-vessels*, and the arrangement of the *valves*, rendered reasonable to the great physiologist (Harvey) the fact of the *circulation of the blood.*

In any case, the fact that the theory of our " future life " is compatible with all the promptings of our religious, moral, imaginative and poetical faculties, that it cannot but tend to elevate and purify our lives, and render them less selfish, is a sufficiently reasonable and *practical* argument, in the absence of all *scientific* proof to the contrary, why this theory, coinciding as it does with the doctrines of Christian Revelation, should be adopted as the basis of our practical rule of life. Be this, however, as it may, the writer having frankly, and he trusts respectfully, expressed his opinions on the subject of the "future existence of man," hopes that he will not be deemed responsible for the action of those who, with a very limited knowledge of "physiology," and a still more limited knowledge of things in general, may attempt to *extort*, or rather *distort*, from statements in this little work arguments subversive of this theory, or of the belief in the existence of the human soul.

4. Comparison of the Actions of a Living Body with those of a Steam Engine continued.—What, then, is a steam engine? and in what respects, it may be asked, can two such apparently dissimilar objects as a *steam engine* and the *body* of a *living man* resemble each other?

A *steam engine* is a complex structure, by means of which—

1st. *Chemical force*, stored up in certain materials (the *fuel*), is converted into *heat* force.

2nd. *Heat force* is converted into *mechanical force*.

3rd. The *mechanical force* is made to do the work desired by the designer and constructor of the engine.

The *real force* which drives the steam engine is *not* the *steam*, but the *heat*. The steam is merely the medium or agent by means of which the heat is most conveniently brought to bear upon the *solid* parts of its machinery.

5. For the construction of a steam engine we require—

(1.) A *grate* or *furnace*, in which the *fuel* is burnt or *oxidized*—that is, made to combine with the oxygen of the air, so that its *chemical force* may be converted into *heat force*. The grate or furnace corresponds to the *lungs* and *capillaries*.

(2.) A *boiler* containing water, placed where it shall receive the utmost possible heat from the *combustion* taking place in the *furnace*. All *heat* which does not pass into the *water* in the boiler is so much *lost force*, and the *fuel* consumed in generating it is so much *lost fuel*.

(3.) A *cylinder*, *piston*, and *piston-rod*, which correspond in function with those of the *bones, ligaments, muscles,* and *tendons* of the living body.

The *cylinder* is a large hollow cylindrical vessel connected with the boiler by a steam pipe.

The *piston* is a circular flat disc of metal, fitting airtight into, and capable of being pushed *up* and down, the cylinder.

The *piston-rod* is a strong metallic rod attached to the *upper* or *outer* part of the *piston.*

When the *water* in the boiler has received the necessary quantity of heat, it expands with enormous force to nearly 1,700 times its original bulk or volume. The capacity of the boiler being fixed and limited, it is unable to contain this additional *bulk* or *volume* of expanded water (steam) which thus *forces* its way, or rather is *driven by the heat* into the *cylinder,* pushing the *piston* and *piston-rod* before it, thus producing the available *mechanical force* for the development of which the steam engine has been set up.

(4.) The *steam governor,* which is a mechanical arrangement by which the quantity of *working steam*—that is, the quantity of steam delivered into the cylinder—is *regulated* according to the resistance of the engine or the *work to be done.* That is, if *more power* is required because *more work* is to be done, then more *working steam* is admitted to drive the piston in the cylinder. If work is taken off so that *less work* is to be done, less *working steam* is admitted,—the fuel consumed being regulated accordingly by the man in charge of the engine.

The *steam governor* of the steam engine, together with the *stoker* or man who attends to the furnace, corresponds to the *sympathetic* or *organic nervous system* of the living man, which regulates or duly adjusts the action of the *digestive, respiratory, circulatory,* and *muscular systems* with regard to each other; so that in health they should not work *too fast* or *too slow* for each other—that is, that the organs of digestion should not make too much or too little blood, or the heart drive it too slowly or too rapidly through the system, or the lungs oxidize it too highly or too feebly, or the tissues appropriate it too quickly.

(5.) *Various mechanical contrivances*, not here necessary to describe, by means of which the *mechanical force*, originating as previously explained in the first instance out of the *heat*, is *distributed* and *directed* so as to accomplish the various ends for which the engine was designed. These parts also, like the piston, piston-rod, &c., thus used for *transmitting* and *directing* mechanical force and movement, may also be compared with the *bones, muscles, ligaments*, and *tendons* of the living body.

6. The amount of work expressed in *mechanical units* done by a well constructed and well managed steam engine (allowing for loss by friction, &c.) is in the ratio of the *fuel consumed*. That is, roughly speaking, for *twice* the *fuel* consumed, *twice* the *mechanical power* should be developed, and *twice* the *work* be capable of being done.

7. According to Helmholtz, the celebrated German physicist, the total *internal* mechanical work of a living man is not less than about 715,000 foot pounds per day, of which 500,500 foot pounds are expended in the *mechanical* work of the *circulation*, 78,650 foot pounds in carrying on the *mechanical* work of *respiration*, and 135,850 foot pounds in the performance of the *mechanical* work of other *internal* processes of the living body.

8. Helmholtz also estimates the *external* work—that is, the *external* resistance overcome by an active, vigorous *working* man per day—at about 715,000 foot pounds.

9. Every $2\frac{1}{4}$ lbs. of good coal burnt in feeding a good steam engine should be capable of producing about 2,145,000 foot pounds of mechanical force; thus the mechanical work producible by the combustion of one kilogramme, or about $2\frac{1}{4}$ lbs. of coal, equals the mechanical work of *three* men for a whole day.

10. Dr. Frankland has estimated that $20\frac{1}{2}$ ounces of *oatmeal* (at a cost of $3\frac{1}{2}$ pence), oxidized in the body of a small living man (weighing 140 lbs.), would enable him

to raise himself 10,000 feet high (as in ascending a mountain); or, in other words, would enable him to expend 1,400,000 foot pounds of *mechanical force*.

11. The steam engine is thus an instrument by which *heat* and *mechanical force* are developed out of the *chemical force* stored up in the *fuel* consumed in working it.

12. **Difference between the Living Body and a Working Steam Engine.**—The *living human body* is, however, a far higher and more complex structure, in which a certain amount of chemical material (the food) is daily *consumed* and *oxidized*, the chemical force evolved during the combustion of which is not simply converted into the ordinary forms of *heat* and *mechanical* (muscular) *force*, but *also* into the various forms of *nervous, vital,* and *mental force*.

13. In the *living body*, as in the case of the *steam engine*, the *quantity of force* developed depends on the quantity of food profitably consumed (that is digested, assimilated, and oxidized). Where more food is *consumed* more force *is developed*. And where *more work*, either *brain* or *muscular*, is to be done, *more food* must be consumed to supply the requisite force.

14. **The Living Body Self-Renewing and Self-Repairing.** But the *living human body*, viewed as a mere machine, *differs* from that of a steam engine, not merely in the greater number and higher nature of the forces developed by the *oxidization* of its *food* and *tissues*, but also in its *self-reparative* and its *reproductive power*.

The steam engine *works* and *wastes* or *wears* itself away, and, therefore, soon requires *repair* from external sources, or *mending* by external agents.

The *living human body*, on the contrary, requires no such external aid, but during health is *self*-reparative, constantly wasting away, constantly expending both *force* and *substance*, yet neither losing power nor weight, that is, preserves from day to day its *average weight* and *strength*.

15. This power of *self-repair* is characteristic only of animate beings. No construction of man, however in-

genious, has ever possessed this property. No man has yet *designed*, much less *constructed*, a clock or other instrument which though always at work, yet should never wear out, or which should *repair* itself or *renew* its parts as rapidly as they wear away; yet this is what occurs in all cases of healthy animal life.

16. Change and Waste. The living body is the seat of actions far more numerous and incessant than those of any mere *inanimate* object however complex.

Even during sleep, *chemical, vital, nervous* and *mechanical* movements are continually proceeding. The blood is ever *circulating*, the heart ever *beating*, the arteries ever *pulsating*, the chest ever *throbbing*, the oxygen of the air "the sweeper of the living organism, but the lord of the dead body," as it has eloquently been described by Professor Huxley, is ever *combining* with and *destroying* the tissues of the body. In fact, during life there is continuous action with its corollary—*constant waste*.

The explanation of the nature and extent to which these processes of continual *change* and *waste* are carried on in the living body forms one of the leading objects of the science of "Animal Physiology."

17. Proof of the development of Heat Force in the living body. Every boy who has played at snow-ball knows that, if he grasp a piece of ice or a small quantity of snow at freezing temperature (0° Centigrade) in the palm of his hand, it will speedily *melt*. He also knows that the *heat*—the *force* by which this effect of liquifaction is produced—is derived from the body. Heat, therefore, is normally *developed* in the *living* body.

What takes place with respect to *ice* held in the hand, would take place with regard to *ice* placed in contact with any other part of the *living* body. Supposing the entire body to be encased in *ice* at *freezing point* (0° Centigrade), and the melted ice (the liquid) to be carefully and accurately *weighed*, the *quantity of heat* given out from the surface of the body might be ascertained with great precision. Physicists thus ascertain by a process of *calorimetry* the *quantity* of heat contained in a body of given weight and temperature.

18. But, in addition to the *heat* given off from the *surface* of the body, a large quantity of the heat developed in the *living* body is also carried off by the *breath*, which as all know is considerably warmer than the air of cold or temperate climates. A knowledge of the origin of "*Animal heat*," and the conditions under which it is developed and distributed through the system, is also included in the science of Animal Physiology, which thus includes an elementary knowledge of chemistry, and of "*heat*," as a branch of "Physics."

The principal *organs* concerned in the development and regulation of the *animal heat* are the *lungs, capillaries*, and the *skin*.

19. **Proof of the development of Mechanical Force in the living body.** Throw a stone up through the air by means of the hand and arm. The *movement* of the stone is a *mechanical* movement, and the *mechanical* force by which it is effected is developed within the living body. At every step we take in walking, we *raise* the whole *weight* of the body ; every time we raise a limb or lift a weight, we *generate* and *expend* mechanical force. *Mechanical force* must therefore be abundantly developed in the bodies of living men.

20. The *development* and *direction* of the mechanical force generated by living beings is effected, as previously stated, mainly by *organs* termed respectively *muscles, tendons, bones* and *ligaments*. The branch of Physiology which specially treats on this subject is termed "*Animal Mechanics*."

21. **Special proof that vital and nervous force are developed** in the body of a living man is quite unnecessary to every thinking man. That *vital, nervous*, and mental force are *generated* or developed in the body of living man is proved by every sensation, thought, and *emotion* of which he is conscious. That every such mental phenomenon is due to the *nervous force* generated by the

chemical action of the *oxygen* of the air on the blood or the tissues—the special organs concerned in producing these phenomena—and the conditions under which they are produced form together so many *natural problems* which it is the object of Human Physiology to solve.

This branch of the study is embraced under the head of the *nervous system.* The chief organs of the nervous *system* are the *brain,* the *spinal cord,* the *cerebro-spinal* nerves, and the *sympathetic* nerves.

22. Waste Processes. But in addition to the *force* developed and expended or *lost, substance* is, in the production of this force, being incessantly *burnt,* disintegrated, and evolved from the *living* animal body, thus producing the continued *loss of substance* or *waste* previously referred to.

To remove this *waste* or *dead* matter with sufficient rapidity from the system is the office or function of special *organs* termed the *absorbents,* and of the organs of *excretion,* the latter including the *lungs, skin,* and *kidneys,* &c.; the duty of which has been compared with that of the *sewers* and *scavengers* in properly organized large towns.

The *waste* substance of the tissues leaves the system in form of *carbonic acid gas,* of *water,* and of *urea;* the latter substance (*urea*) after leaving the body becomes further decomposed into *carbonic acid gas* and *ammonia.* In addition to the above, the *excretions* also contain certain *saline compounds* or salts.

23. Simple Proofs of Waste.

(*a.*) Bring any bright highly-polished cold steel or metal article into contact with the finger or any other part of the body. It will immediately become *dimmed,* because of the deposit of the *perspiration* which incessantly escapes from all parts of the *skin.* Ten thousand to twenty thousand grains weight and upwards are thus thrown off the body daily.

(*b.*) Breathe through a *weighed* quantity of clear transparent *lime water* (a chemical test for *carbonic acid gas*). It instantly becomes *white* and *turbid.* After breathing

14 E.   B

through it for a short time, *re-weigh;* it will now have become *heavier;* thus proving that the living body continually *loses carbonic acid gas* in the act of breathing. A large quantity of *aqueous vapour* also passes off from the lungs during the process of breathing.

Twelve thousand to twenty thousand grains in weight of *carbonic acid gas,* and five thousand grains and upwards of *water* in the form of *aqueous vapour,* are thus lost per day from the body of living man.

During *cold* weather thick *clouds* of condensed vapour are frequently seen rising from the *mouths* and *skins* of horses who have been running violently; this is simply so much of the *waste matter* referred to.

24. Waste increases with Work.—Let a living man be *weighed* immediately after a meal; let him then sit still for three or four hours, after which let him be *re-weighed,* he *not* having taken any refreshment in the interval. On *re-weighing* he will be found to have *lost weight,* thus proving that during the interval he was *losing substance.*

Let the same experiment be repeated after another meal, the temperature being the same, but let him in the interval *work* hard, or take a very long, rapid walk. On being *re-weighed* he will be found to have *lost* a very perceptibly *greater* weight than on the occasion of the previous experiment during *rest;* thus proving that *increased exertion* brings with it *increased loss* of substance.

A smith or a navvie thus on ordinary days, when *working,* loses more bodily weight and substance than he does on Sundays when *resting.*

25. Diurnal balance.—But if a healthy living man be *weighed* at about the same time of the day, and under similar circumstances with regard to *temperature, work* and *food,* day after day for several days in succession, he will be found to vary very little in weight from one day to the other. The body must thus be "*diurnally balanced,*" and this *diurnal balancing* must therefore be effected by the food *diurnally* ingested.

The body of a living man is thus a highly complex *organized* structure, which constantly *generates* and *expends* nervous and mechanical force, which *generates* and *loses* heat, and which consequently suffers an *incessant loss* of substance *consumed in supplying* the forces thus generated and expended, and which must therefore be sustained by the periodical supply of suitable *organic* matter ingested in the form of *food.*

If a healthy active man do a given amount of physical work daily, he will require a certain definite amount of food per day to preserve his *diurnal physiological balance.* If he *greatly* increase the quantity of *work* done per day without increasing the quantity of *food* he takes per day, he will become *thinner* and *lighter*—that is, his diurnal bodily loss will *exceed* his diurnal bodily gain. Again, if he continue to take the same quantity of food as before, while he greatly reduces the quantity of work he does per day, either his body will become *heavier*—that is, he will increase in *bulk*, or the unused food will pass out of his system *undigested* as *excrementitious* matter.

26. **Hunger and Thirst.**—In order to compel the *living* being to attend with sufficient promptness and regularity to the supply of new matter in the form of food and drink necessary to keep up the *diurnal balance*, and enable the body to generate the *forces* necessary to the carrying on of its functions, and the repairing of its tissues, *two imperious* sensations, *hunger* and *thirst*, are established, which, when operating in their full vigour, irresistibly compel him (where physically possible) to satisfy the cravings of his system. So irresistible are these cravings that savages sometimes charge their stomachs with *clay* and other indigestible and useless matter in order to assuage the intensity of their *hunger.*

27. **The Ingestion of Oxygen a condition of Life.**—But as the development of the *vital* and other *forces* generated in the living body depends upon the chemical action of the *oxygen* of the air upon the *blood*, the *food*, or the *tissues*, it is necessary that, in addition to the food

material ingested for the *repair* of the system, large quantities of *oxygen gas* shall be constantly ingested into the system.

It has been estimated that 800 lbs. of *oxygen gas*, and consequently about five times that weight of atmospheric air are passed through the *lungs* of an ordinary working man in the course of one year. (See *Organs of Respiration.*)

28. When the body ceases to be supplied with *oxygen*, the *brain*, the *heart*, and the *lungs* cease to act, and death from *asphyxia* or *suffocation* ensues.

29. **Death, Local, Molecular, and Somatic.**—When the body of an animal *performs* its various *functions* it is said to *live*, or be in a state of *life;* when such a body entirely *ceases* to perform its various functions, it is said to be *dead.*

Every animal is endowed at its birth with a constitution, in consequence of which it is, under favourable, that is *normal*, circumstances, capable of passing through a systematic series of cycles of change at the termination of which it, *no longer possessing the power* of continuing to develop the *vital* forces necessary to its further existence, *ceases to live*—or in other words, dies a *natural*, and probably *painless* death, from pure natural exhaustion of vitality.

This constitutes death from *old age.* From misconduct, and breach of natural law, in most cases more or less unavoidable, because of ignorance of physiological law, or of the artificial necessities of modern civilization, probably not one in ten-thousand *dies a natural death*—that is, lives out the full period of his proper natural existence, but *dies* by accident (injury) or by *disease.*

Death, however, does not occur *simultaneously* through every part of the body. The *tissues* of the man continue to live, and even to be nourished for a short time after the *man* himself is dead. Thus the *hair* may possibly continue to grow a short time after death. The *muscles* also may be made to contract by *electrical stimulus* for a short time after the man is dead; thus showing that the *muscular tissues* are not dead.

When the *abdomen* of a sheep which has been bled to death, or even decapitated, is opened shortly after death, the *peristaltic* movements of the stomach and intestines, probably stimulated by the action of the air, may be observed proceeding feebly; thus again showing that the *life* of the *tissues* may for a short time *survive* that of the body as a whole.

The *death* of the body as a whole has been termed *Somatic death* (from Greek *soma*, a body). *Somatic death*, formerly described as *systemic death*, is death consequent on the *cessation of the circulation*. The cessation of the circulation can only be brought about by the *failure* of the action of the *brain*, the *heart*, or the *lungs*. These three organs, or centres of life, were therefore designated by *Bichat* the *tripod of life*.

When, as not unfrequently happens, a part of the body, as a finger or a limb, suffers injury by accident or disease, in consequence of which its circulation and consequent *nutrition* is arrested, it *dies*, or undergoes *mortification*, and sloughs away. Such death is therefore termed *local death*, which implies the death of a *part* of the body in contradistinction to the death of the whole. (See *Nutrition.*) Birds *moult* their feathers, and deer *cast* their antlers through *local death*, caused by arrest of nutrition.

But it has already been repeatedly shown that the *vital* and *other forces* of the body—or in other words, *its life*, is continually sustained by the disintegration, oxidizing, burning, and consequent *death* of the *molecules* (minute constituent particles) of the tissues taking place *at all points* in the *living* body. To this kind of *death*, therefore, the term *molecular death* is applied, which implies that kind of death which is perpetually taking place among the living *particles* all through the body, the death and destruction of which (through the agency of the oxygen) determines the origin of the *animal* heat, and of the muscular, nervous, and other forces of life.

Molecular death taking place all through the system has sometimes been described as one kind of *general death*, and has thus been confounded with *somatic death*, which is also another form of *general* death.

30. Reproduction.—To the power of reproduction, so characteristic of animal life, by which life is given to successive races of beings, who inherit the *structure* and *properties* of their *parents* and predecessors, the limits of the present treatise will prevent little more than allusion.

31. Cognate Sciences.—It will be seen from the foregoing statements that a sound knowledge of Animal Physiology implies a greater or less knowledge of *Chemistry, Mechanics*, and *Physics*. It also requires a knowledge of so much *Anatomy* as shall enable the student to understand the general *structure* of the Animal Body, and of so much *Histology* as shall give him a clear knowledge of the *Microscopic structure* of the tissues. Most important light is thrown on the principles of physiology by the study of *Pathology*.

CHAPTER II

GENERAL BUILD OF THE HUMAN BODY.

32. Divisions of the Human Body.—For the purposes of general description, the human body may be divided into *head, trunk*, and *extremities* (the arms and legs). The human body is, speaking generally, *bi-laterally symmetrical*—that is, it consists of two similarly shaped and equal halves, right and left, each of which is made up of similar parts or *organs*.

Before entering on the more *minute* study of the "house we live in," it is desirable to take "a run of the house;" the student should therefore endeavour, in the first instance, to obtain a clear idea of the "General Build of the Human Body."

33. The Head, which forms the upper part of the *body*, contains—the *brain* or organ of *thought, sensation*, and of the *emotions*—the organs of the *chief senses*—viz., those of sight, smell, taste, and hearing, and the *medulla oblongata,* or cranial portion of the central nervous axis. (See fig. 1.)

That portion of the *head* which contains the brain is termed the *cranium,* the remaining portion, the *face.* The

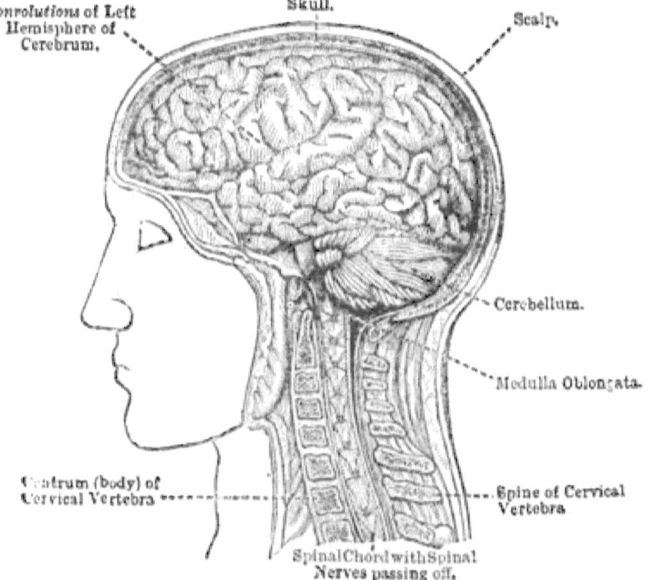

Convolutions of Left Hemisphere of Cerebrum.

Skull.

Scalp.

Cerebellum.

Medulla Oblongata.

Centrum (body) of Cervical Vertebra

Spine of Cervical Vertebra

SpinalChord with Spinal Nerves passing off.

Fig. 1. Side view of Brain and connection of Spinal Cord.

mouth and nostrils open into the *pharynx;* the ducts from the *salivary glands* open into the mouth.

34. **The Trunk,** which forms the large mass of the *body,* may, in order to facilitate description, be further divided into an *upper* part, termed the *thorax* or chest, and a *lower* region, termed the *abdomen.* (See fig. 2.) The *thoracic* and *abdominal* cavities are separated from each other by a large thin *muscular* partition, termed the *diaphragm,* or midriff.

35. **The Thorax** or chest contains the *thoracic cavity,* in which are lodged the heart, lungs, trachea, and portions of several of the larger blood-vessels (the aorta, vena cava, and pulmonary vessels), and the *dorsal* portions of the *spinal cord,* and of the *bony axis* by which it is protected, also a portion of the *œsophagus* (gullet or food-pipe), and

the *thoracic auct,* and of the *sympathetic* or ganglionic-nerve-system, not shown in the diagram.

The *spinal cord,* which is a sort of continuation of the *brain* down the middle of the back-bone, transmits nervous impressions *to* and mental commands *from* the brain to the various parts of the body. It also acts as an *independent* centre of motion or *reflex* action.

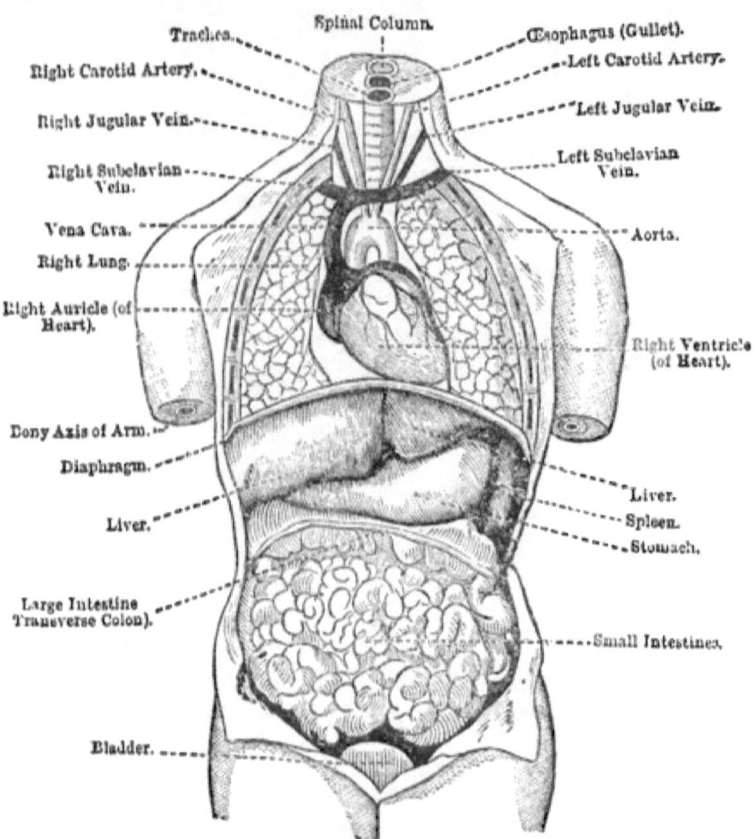

Fig. 2. Front view of the Organs of the Trunk.

The *walls* of the thorax are strengthened for the more secure protection of their *visceral* contents by a bony cage-work, consisting of the ribs, the dorsal portion of the back-bone, and the breast-bone.   The *sternum* (breast-bone) and the front portions of the ribs (the *cut ends* of which are seen) are supposed to be removed in fig. 2, in order to expose the contained organs.

The *cavity* of the thorax contains the chief *blood-purifying* and *blood-circulating* organs.

36. **Transverse Section of the Thorax.**—The following diagram sufficiently explains the structure of the *thorax* and *its contents*, as displayed by a section across the *heart* and *lungs* perpendicular to the vertebral column (back-bone), the outer integument and layer of subcutaneous fat having been removed.

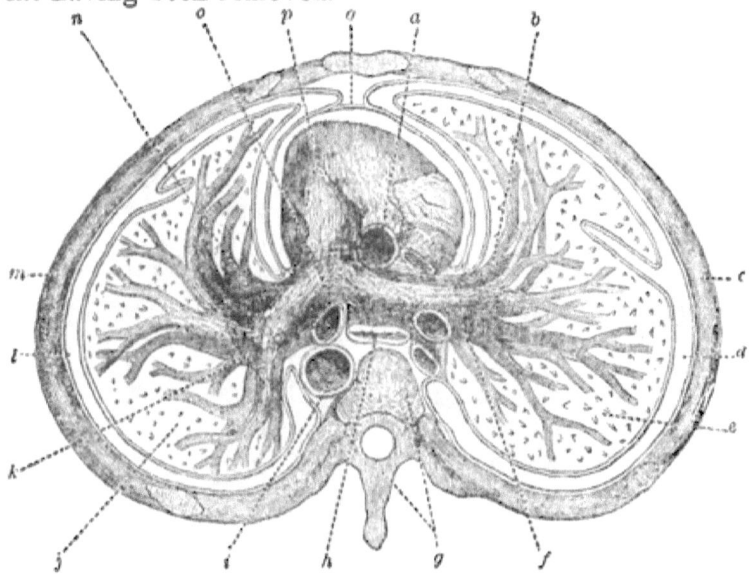

Fig 3.   Transverse Section of the Thorax.

a, Aorta (ascending). *b*, Right Pulmonary Vein. *c*, Right Rib. *d*, Interval between two Surfaces of Right Pleura. *e*, Right Lung. *f*, Right Bronchus *g*, Vertebra. *h*, Œsophagus *i*, Aorta (descending). *j*, Left Lung. *k*, Left Bronchus. *l*, Left Pleura. *m*, Left Rib. *n*, Left Pulmonary Vein. *o*, Pulmonary Artery. *p*, Heart. *q*, Pericardium.

37. **The Cavity of the Abdomen,** situated immediately *below* the *diaphragm*, contains the *liver*, the *stomach*, the large and small *intestines*, and the *bladder*, also the *lumbar* portion of the back-bone and spinal cord, the *pancreas*, the *spleen*, and the *kidneys*, not shown in fig. 1.   The cavity

of the abdomen thus contains the principal *blood-forming* organs.

38. Hints for Lay Students of Physiology.—The young student should learn quickly to distinguish between the terms *body* and *trunk*. He should also make himself thoroughly familiar with the *forms* of the various organs, and the *positions* they occupy in the *living* body. He should study and draw the various physiological diagrams just as he would study a geographical map, and should always, where possible, avail himself of the opportunity of studying the actual organs they represent in the bodies of animals displayed by the butcher in his shop. In all cases he must make himself thoroughly *familiar* with each subject, especially with its *nomenclature* (technical names), before he proceeds to the study of the next, or he will make his work *needlessly difficult*. He should also bear in mind that after once carefully reading the description of a particular organ, observation or examination of the organ itself will give more *real* knowledge than many hours of laborious reading. He should also bear in mind that in the bodies of the sheep, or of the pig, or even the rabbit (commodities of common food), he has the means of securing nearly all the opportunities for accurate observation required for the sound study of animal physiology.

Of course, the structure of the bodies of these animals differs in many respects from that of the human body; notwithstanding this, it is easy, with the aid their study affords, supplemented with that of diagrams, to acquire in a most interesting manner a very sound knowledge of the *functions* of the *human body*, and of the general principles of *Animal Physiology*.

39. The Extremities or limbs are attached by an *upper* and a *lower girdle* of bones, externally, to the trunk. They consist essentially of *solid* fleshy or *muscular* (contractile) organs, and contain no *cavities* similar to those of the head and trunk.

The *upper* limbs or arms, which are terminated by the hands, possess great power of mobility—they are chiefly organs of *motion* and *prehension*. Their structure will be explained more fully in describing the *skeleton* or bony framework of the system.

The *lower* limbs or legs, which are terminated by the feet, are, in the human being, much larger and more powerfully built than the *upper* limbs. Their chief *functions* are those of *support* and *locomotion*.

The greater length, size, and strength of the lower limbs of man afford so many anatomical and physiological proofs that the *erect* position of man during walking is his *natural*, and not, as has been insinuated, his merely *acquired* position.

40. The terminal expansions (the feet) of the lower limbs resemble, in *general plan* and *structure*, those of the hands, but differ chiefly in being *less mobile* and less perfect as *sensory* and *prehensile* organs. Their less prehensile power mainly results from the fact that the bones of the great toes are not *opposable* to the bones of the remaining toes or of the instep; whereas the bones of the thumbs are readily opposable to the bones of the other fingers or to the bones of the palms of the hands.

Though in general the degree of manipulative power and skill possessed by the feet is many degrees inferior to that possessed by the hands, yet the effect of practice and training in augmenting the power of the former is most wonderful. The writer recollects, on one occasion in the picture gallery at Brussels, being astounded at the facility of execution and power of handling the brush and the pencil shown by one of the artists then painting in the gallery. This gentleman, who was entirely deficient of *both* arms, executed most charming copies of the leading works in the gallery, "handling" his panels, his canvas, his pencils, and his pigments, exclusively by means of his legs, feet, and toes.

41. The *limbs* consist essentially of *bony axes*, surrounded by large masses or bundles of *fleshy fibre*, by which they are moved. They also contain *whitish, silky*-looking cords or threads, the *nerves*—the "telegraph wires"—by which the command and the power to move are transmitted, or as it were telegraphed, to the muscle, which thus becomes the obedient servant of the *brain*.

The limbs are also supplied with large blood-vessels (arteries and veins), by which they are furnished with the blood necessary for nutrition or *self-repair*, which has been previously described as the grand triumph of *living* over *art structure*.

The large *blood-vessels* and the principal *nerves* of a limb take the direction of, and are situated for protection, near its bony axis.

The general structure of a limb may be well studied by means of the leg of a fowl or a rabbit, even after it has been served up at the dinner table. The *tendons*, the *ligaments*, and the *fasciæ*, however, become more or less gelatinous during the process of cooking.

42. Transverse Section of a Limb.—If a *limb*, say the leg, were cut *transversely* through its middle—that is, perpendicularly to its general length—the student would observe, commencing from its exterior—

a. A thin circular coating or integument, the *skin*.

b. A circular layer of fat (the layer of subcutaneous fat).

c. A large mass of red flesh, consisting of *bundles* of *muscular fibre*, each bundle having its own coat or sheath, the whole mass also being surrounded by its own smooth shining sheath or *fascia*.

d. A central hollow, more or less cylindrical, bony axis, the interior being either empty or filled with medullary matter (marrow).

e. Two or three large blood-vessels and nerve-trunks, situated close to the bony axis.

All the structures here indicated may be readily seen in an ordinary leg of mutton, as sold by the butcher.

43. Investing Membranes.—The *exterior* of the body is surrounded by the skin or outer integument, which consists of *two* layers—an outer *bloodless* and an inner *vascular* and *sensory* layer.

At the mouth the skin enters the *interior* of the body, and its outer layer undergoing some modifications, it becomes *mucous* membrane, which lines the *open* cavities in the body.

The *closed* cavities are lined by *serous* membranes —so called from the fluid which they secrete and by which they are moistened.

44. Blood-Vessels, Nerves, Absorbents, and Glands.—For the *general course and distribution* of the blood-vessels, the nerves, the *absorbents* or *lymphatics* (organs which absorb or remove from the system the used up, or partially used up, materials from the body), and the various *secretory* and *excretory glands* (organs by which substances are elaborated and impurities are eliminated from the blood), the student is referred to the *special diagrams* illustrating these portions of the system.

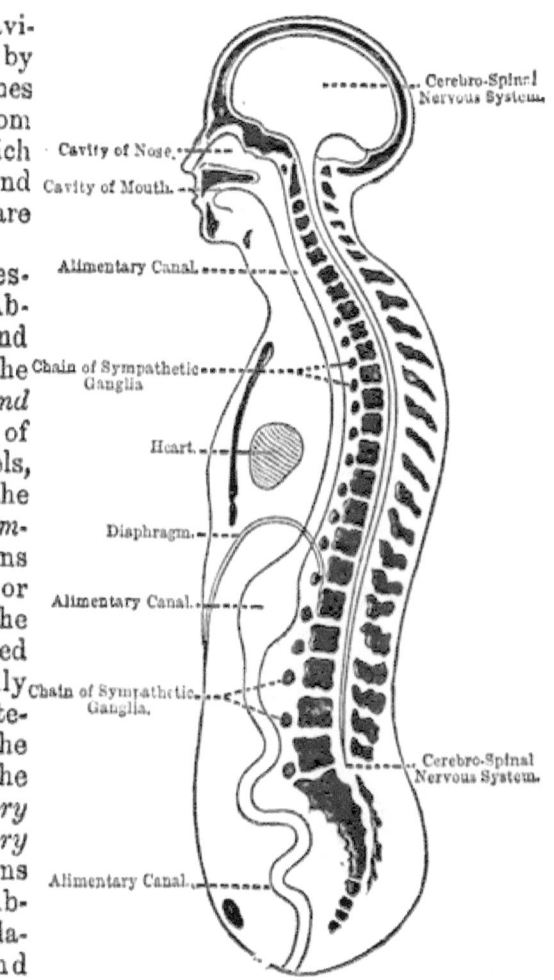

Cerebro-Spinal Nervous System.

Cavity of Nose.

Cavity of Mouth.

Alimentary Canal.

Chain of Sympathetic Ganglia

Heart.

Diaphragm.

Alimentary Canal.

Chain of Sympathetic Ganglia.

Cerebro-Spinal Nervous System.

Alimentary Canal.

Fig. 4. Theoretic Longitudinal Section of Human Body, showing Dorsal and Ventral Tube. (After Huxley.)

The section is taken *perpendicularly* through the *Median* plane, and shows the *dorsal* (neural) tube containing the dorsal chamber of the skull and the spinal canal, the cut surfaces of the skull and 33 vertebræ, and the *ventral* (hæmal) tube in front of the Vertebral Column, containing the Heart, Lungs, Alimentary Canal with cavities of Mouth and Nose, the *Sympathetic Nervous System* consisting of a *double chain of Ganglia*, and other organs.

45. Double-tube Theory of Structure of Animal Body.—

According to the theory now generally adopted by physiologists, the *body* of a *vertebrate* (back-boned) animal consists essentially of a *double-tube*, the *walls* of the tubes being united, but their *cavities* separated by the bodies of the vertebræ.

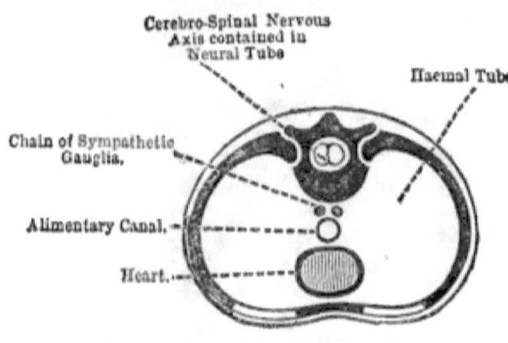

Cerebro-Spinal Nervous Axis contained in Neural Tube

Hæmal Tube.

Chain of Sympathetic Ganglia.

Alimentary Canal.

Heart.

Fig. 5. Theoretic Transverse Section of the Human Body showing Dorsal and Ventral Tube. (After Huxley.)

The *posterior* or *upper* tube is termed the *dorsal* or *neural* tube or canal. It contains the *brain* and *spinal cord* (the cerebro-spinal nervous system).

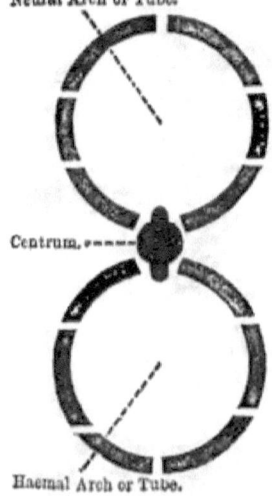

Neural Arch or Tube.

Centrum.

Hæmal Arch or Tube.

Fig. 6. Typical Vertebra.

The different Segments of the arches represent (typically) various extensions and processes of the Vertebra.

The anterior or lower tube or canal is termed the *ventral* or *hæmal* tube. The *hæmal* or *ventral* canal includes the face from the eyes downwards (the mouth and nose forming a double inlet), the *heart, lungs, alimentary canal*, and other *blood-preparing* organs, and the *sympathetic nervous system*, and other organs.

The *double-tube* is made up of a series of *segments*, termed *vertebræ*, which are built up or super-posed, the one on top of the other.

46. The *theoretical typical vertebra* (see fig. 6) is supposed to consist of *two* bony arches, rings or hoops, connected by a *body* or *centrum*. The one is intended to contain a portion of the nervous system, and

is therefore termed the *neural* hoop or arch (from Gr. *neuron*, nerve); the lower hoop is intended for the protection of a portion of the *vascular* system, and is therefore termed the *hæmal* hoop (from Gr. *haima*, blood).

47. It is further supposed that the *skull* consists of four greatly modified vertebræ, in which the *neural* (dorsal) *arches* or hoops are greatly enlarged; also, that the abdominal and thoracic cavities are more or less enclosed by *vertebræ*, greatly *modified* by the addition of ribs (pleurapophyses), &c., which are regarded as mere extensions of the *hæmal* or *ventral* hoops. The *pelvis*, which bounds the lower end of the *ventral* tube, is also regarded as consisting of *modified* vertebræ.

CHAPTER III.

THE SKELETON OR OSSEOUS SYSTEM—THE BONES AND LIGAMENTS.

48. The Skeleton (from Gr. *skello*, I dry up) is the hard bony framework of the body. It consists of 200 or upwards of separate *bones*, united together by means of *cartilages* and *ligaments*. It, like the body, consists of *head*, *trunk*, and *extremities*. (See fig. 7.)

The difference in the number of the bones, as estimated by different writers, arises from the fact, that many of the bones are compound, consisting of several parts, which in early life are quite distinct from each other, but which later in life become more or less connected, so as to form single bones.

The student should make himself thoroughly familiar with the general plan and structure of the *skeleton*, and with the *names, positions*, and *shapes* of its various bones, since the general direction, and the *names* of the bones, determine the direction and the names of a large number of the blood-vessels, nerves, muscles, &c., to which they are adjacent. The whole of this may be *well* and *pleasantly* taught to a class of students, of from twelve or thirteen years of age or upwards, with the aid of a

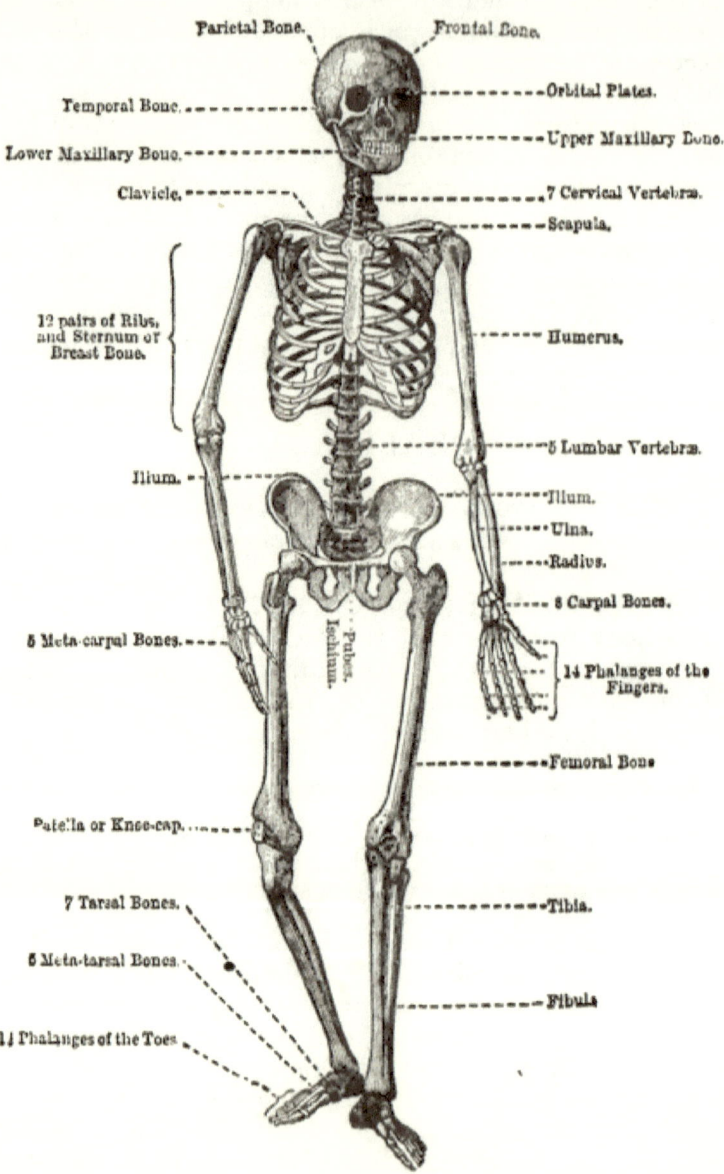

Parietal Bone.

Frontal Bone.

Orbital Plates.

Temporal Bone.

Upper Maxillary Bone.

Lower Maxillary Bone.

Clavicle.

7 Cervical Vertebræ.

Scapula.

12 pairs of Ribs, and Sternum or Breast Bone.

Humerus.

5 Lumbar Vertebræ.

Ilium.

Ilium.

Ulna.

Radius.

8 Carpal Bones.

5 Meta-carpal Bones.

Pubes.
Ischium.

14 Phalanges of the Fingers.

Femoral Bone

Patella or Knee-cap.

7 Tarsal Bones.

Tibia.

5 Meta-tarsal Bones.

11 Phalanges of the Toes

Fibula

Fig. 7.

black-board and *skeleton*, by a good teacher, in the course of *two* lessons of less than an hour each.

The thirty-two *teeth* belong to the *tegumentary* (skin) system, and not to the *osseous* system; that is, they form no part of the *skeleton* proper.

49. Properties of Bone.—*Bone*, as we usually see it *out of the body*, is a solid, hard, yellowish-white, inflexible, tough, durable, and, compared with its strength, *light* substance. (See *Osseous Tissuë*.)

50. Living Bone, when exposed, as by wound or injury, as it exists in the *living* body, has a *reddish* or *pink* colour, due to the blood circulating in its larger capillaries.

51. Composition of Bone.—Human bone consists of about one-third of *organic* or *animal* matter, and two-thirds of earthy matter. *The animal matter* consists chiefly of *connective tissue*, often improperly termed *gelatine*, because it yields gelatine on boiling. It gives *toughness* to the bones. The *Earthy matter*, which gives *hardness* and *durability* to the bones, consists chiefly of *salts of lime*, chiefly *phosphate of lime* (*calcium phosphate*), and *carbonate of lime* (chalk or *calcium carbonate*), of which there is nearly five times as much *phosphate* as *carbonate*. It also contains small quantities of phosphate of magnesia (magnesium phosphate) and common salt (sodium chloride.)

EXPERIMENT I.—Place a bone on the top of a bright red-hot fire, until all the animal matter has been decomposed or burnt. The *residue*, which consists purely of the *earthy matter*, will remain. It will be extremely *brittle* and *inflexible*, but will retain the *shape* of the bone.

EXPERIMENT II.—Immerse a long bone for a few days in dilute *nitric* or *hydro-chloric acid*. The *earthy matter* will dissolve out, leaving the flexible *matrix* of the bone, consisting of *animal matter* (connective tissue). The bone will now have lost its *hardness* and inflexibility, but it will still retain its toughness, and may now be bent or twisted into a knot.

52. Bones in Infancy and Old Age.—The *quantity* of *earthy matter* in the bones, however, varies greatly at different periods in life. During infancy they scarcely contain any earthy matter, and are said to consist almost entirely of cartilage. At this period the bones are

14 E.   C

comparatively soft and flexible—they consequently *bend* easily, and do not *break*.

53. During *old age*, however, the quantity of *earthy matter* increases very greatly : the bones consequently become very *brittle*; and, *if broken*, in many cases will not again unite.

54. The lower we go in the scale of animal life the less the quantity of *phosphate* of lime, and the greater the quantity of *carbonate* of lime do we find in the skeleton, until at last the former almost entirely disappears from it.

55. **Classification of Bones by Shape.**—Bones are divisible according to shape into *four* classes, viz. :

(1.) *Long Bones*, chiefly found in the limbs, where they form *levers*. The long bones beginning upwards are the clavicle, humerus, radius, ulna, femur, tibia, fibula, metacarpal, and metatarsal bones, and the phalanges.

(2.) *Short bones*, as the carpal and tarsal bones (those of the wrists and ankles.) They consist of an external crust of hard *compact* bony tissue, the whole of the interior of the bone being composed of loose or *cancellated* bony tissue.

(3.) *Flat bones*, as the large bones of the cranium (the frontal, parietal bones, &c.), the ossa innominata sternum, &c. (See *Diploe*.)

(4.) *Irregular bones*, that is bones which cannot be found classified under either of the preceding heads, as the sphenoid and ethmoid bones of the skull, the inferior turbinated bone of the nose, and the hyoid bone of the tongue.

56. **Division and Growth of Long Bones.**—All the long bones consist of a shaft and two extremities.

The *Shaft* or *Cylinder* is a long *hollow* cylinder, the thick walls of which are composed of *compact* bony tissue. The hollow space in the interior, which contains the *marrow*, is termed the *medullary canal*.

The upper extremity of the large bones is termed the *head* of the bone. The two extremities are usually much expanded, frequently forming *condyles* (from Gr. *condulos* knuckle),

The *Long Bones* grow in *thickness* by the deposition of new bony matter in *successive layers*, by the inside of the *periosteum*, or the *outside* of the bone.

The *long bones* grow in *length* from the *ends* of their *shafts*. The *extremities*, termed the *epiphyses* (from Gr. *epi*, upon, and *phuo*, I grow), of the long bones are, until *adult* age, when the bones have ceased to grow, *separated* from the *shaft* by a kind of *cartilaginous* layer, which dips in between the ends of the shaft and the *epiphyses* or extremities. The *growth*, in *length*, of the bone takes place in this cartilaginous layer, chiefly in the *surface* towards the end of the *shaft*.

57. Periosteum (from Gr. *peri*, round; *osteon*, bone). The exterior of the bone, except the parts covered by articular cartilage, is lined by a thin, firm, tough vascular membrane, consisting of white fibrous tissue, termed the *periosteum*. When the *periosteum* of any portion of a bone is seriously injured, *necrosis*, or death of that portion of the bone sets in, because of the interruption of its *nutrition*.

The *periosteum* serves—1st, As a *medium of attachment* to the bone for the *muscles, tendons*, and *ligaments;*

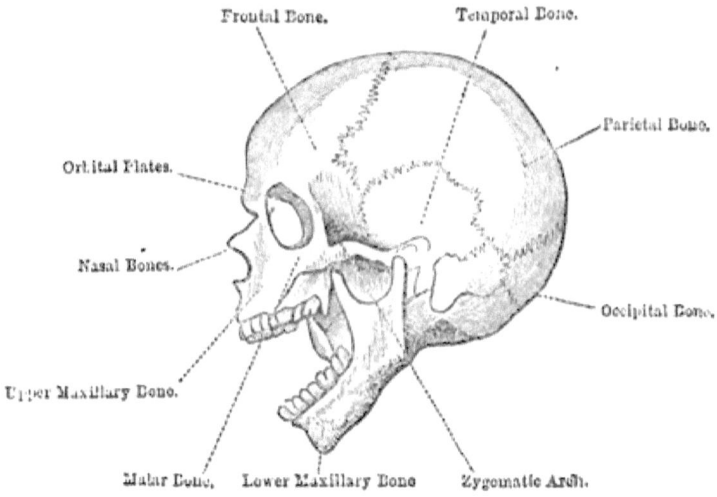

Fig. 8. The Human Skull.

2nd, It *lessens friction* by making the surface of the bone smoother; 3rd, It constitutes a medium, or nidus, in which the blood-vessels intended to *nourish* and *vivify* the bone ramify and break up into smaller branches before they can enter the minute orifices in its surface and be distributed through its substance.

58. **Processes** (from Lat., *pro*, before; and *cedo*, I go).—The various eminences, projections, protuberances, &c., projecting from the surface of the bone for the *attachment* of muscles, tendons, or ligaments are termed *processes;* as the *occipital,* the *odontoid,* and the *mastoid* processes, and the *trochanters* of the *femur.*

59. **The Head,** which is mounted on the *atlas* or topmost bone of the vertebral column, consists of the bones of the *cranium,* and of the *face.*

60. **The Cranium** (from Gr. *kranion,* skull) is the oval bony shell, the *brain-case,* by which the *brain* and *medulla oblongata* are enclosed and protected.

The cranium consists of eight bones, viz.: the *frontal* bone, the two *parietal* bones, the two *temporal* bones (the temples), the *occipital* bone, the *ethmoid,* and the *sphenoid* bones. The two latter bones, not shown in the diagram, are situated at the *base* of the skull, over the back of the root of the nose. (See Fig. 8.)

The *essential* parts of the ears are contained in the *temporal* bones. The eyes are lodged in the *orbits.* The *orbital plates* are formed by the union of the bones of the skull and face.

The skull is regarded by philosophical anatomists as consisting of four *modified vertebræ,* each of which forms part of the *double tube* of bone previously referred to.

61. **The Dura Mater** is the dense, thick, *inelastic* membrane which lines the interior of the skull, and forms the *falx, tentorium,* and *venous sinuses* of the brain.

62. **Diploë** (from Gr. *diploos,* double).—The loose *cancellated* bony tissue represented by the shading in the middle of the bones of the skull, in fig. 1, is termed the *diploë.* The flattish arched bones of the skull consist of two layers or *tables* (an outer and an inner) of hard or *compact* bony tissue connected by a *diplous* or cancellated

layer. ¯ By this arrangement of structure, the bones of the skull are not only *lightened*, but external injury or *fracture* is frequently prevented from passing to the *interior* of the skull. Most of the *flat* bones possess this structure.

63. **Sutures.** (from Lat. *suo*, I sew). The bones of the *cranium* and *face* are joined *immovably* to each other by means of *dove-tailed* or somewhat serrated edges; the bones being presented *edge* to *edge*, the *projections* of the one bone fitting into corresponding *indentations* of its adjacent bone, much in the same way that a cabinet-maker unites the sides of a well-made box or drawer to each other. These *joints*, termed *sutures*, owe their name to their *sewed* or *seam-like* appearance.

The bones of the *cranium* grow from their *edges*, by which they thus adapt themselves to the increasing size of the *growing brain*.

Where the bones do not properly meet edge to edge, but *overlap* each other like the scales of a fish, they are termed *squamous* (from Lat. *squama*, a scale), sutures.

The *sagittal suture* connects the two *parietal* bones, the *coronal suture* connects the *frontal* and *parietal* bones, the *lambdoidal suture* unites the *occipital* with the *parietal* bones, and the *squamous suture* unites the *temporal* with the parietal bones.

64. **The Face** contains fourteen bones, viz. :—two *nasal*, two *upper maxillary* (jaw), one *lower maxillary*, two *malar*, two *palate*, two *lachrymal*, one *vomer* (septum of nose), and two *inferior turbinated* bones. The face contains the *cavities* of the mouth, nose, and eyes, the cavities of the latter being termed the *orbits*, thus together enclosing five cavities, which contain the organs, sight, smell, and taste.

65. **The Hyoid Bone**, or the tongue bone, is the U-shaped bone situated between the *tongue* and the *larynx*, to which the muscles of the tongue are attached.

66. **The Vertebral or Spinal Column** is the long, bony double tube or column, which, giving support to the head, passes down the *median* line at the back of the trunk, and joins the *pelvis* at the lower end of the trunk.

It consists of twenty-four *immovable* vertebræ, the *os sacrum*, which consists of five imperfect *fixed* vertebræ, united into one bone, and the *os coccygis*, which contains four imperfect vertebræ.　The *spinal column* thus contains in all thirty-three vertebræ, consisting of seven *cervical* (neck), twelve *dorsal* (back), five *lumbar* (loins) vertebral bones, together with the *os sacrum* and *os coccygis*, which are all clearly shown in fig. 9.

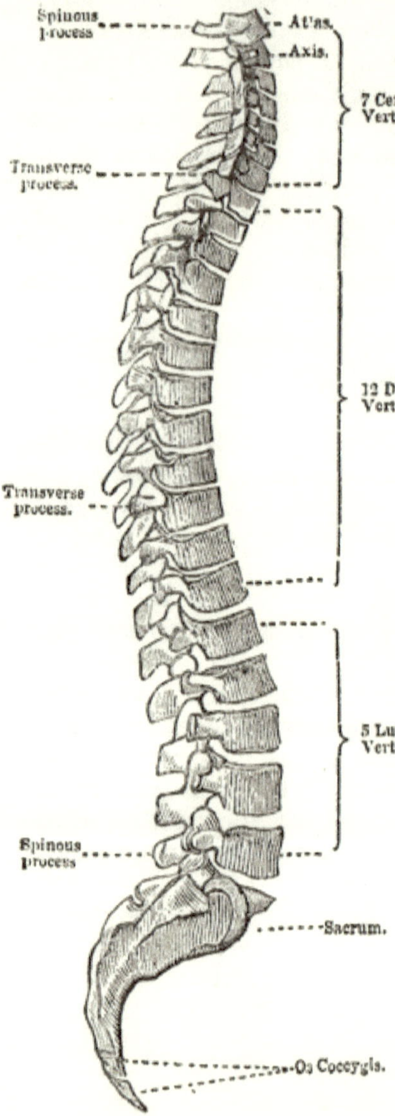

It will be observed on examining the diagram of the vertebral column, that the *vertebræ* increase in size and *strength downward*, because of the greater burden they have to bear, thus affording additional structural proof that the *erect* is the position natural to man.　It will also be observed that the transverse and lateral *processes* become larger, especially at the loins, for the attachment of larger and more powerful muscles.

67. **A Vertebra** (from Lat. *verto*, I turn) is a single complete *segment* of the *vertebral column*, or *bony axis* of the *trunk*.　It is one of the *irregular* bones: its *essential* parts are the *body* or *centrum* (its anterior segment), and a posterior segment, the *arch*.　(See figs. 10, 11.)

Fig. 9.　Vertebral Column.

Spinous Process
Atlas.
Axis.
7 Cervical Vertebræ.
Transverse process.
12 Dorsal Vertebræ.
Transverse process.
5 Lumbar Vertebræ.
Spinous process
Sacrum.
Os Coccygis.

But the *vertebræ* also usually contain *transverse* and *spinous processes* for the attachment of the muscles by which the body is *supported, bent,* and *turned; lateral*

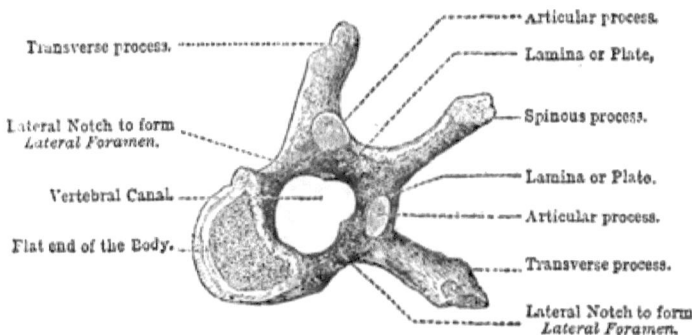

Fig. 10. Top of a Vertebra.

notches by the superposition of which the *intervertebral foramina* (lateral apertures), by which the *spinal nerves*

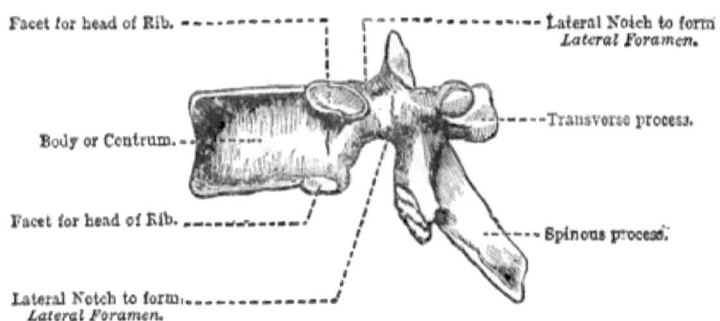

Fig. 11. Side of a Vertebra.

leave the *spinal canal*, are formed—*articular surfaces* and *facets*, by which they are joined to *each other* and to the *ribs.*

The *bodies* or *centrums* are held together, one over the other, by the *intervertebral layers* of *fibro-cartilage*, their arches being connected by the *ligamenta subflava* of

the *vertebral column*.   The *spinal foramen* (hole) *of each*
bone also being superposed, form the *spinal* or *vertebral
canal*, which encloses and protects the *spinal cord.*

By this arrangement, each of the twenty-four movable
vertebræ thus yielding a little, a considerable amount
of *bending* and *torsion* of the *vertebral column* is secured
without injury or compression of the enclosed *spinal
cord*.   If the spinal column were *bent* suddenly and
sharply as in the case of the elbow (a hinge joint), the
spinal cord would be compressed, immediate paralysis
would be produced, and the body would instantly fall.

The bones of the *head* are supposed to consist of *four
modified vertebræ*, as previously explained.

68. **The Pelvis** (Lat. a basin) is the *girdle* of bones at
the lower end of the trunk which supports the contents
of the abdomen, and transmits the weight of the body

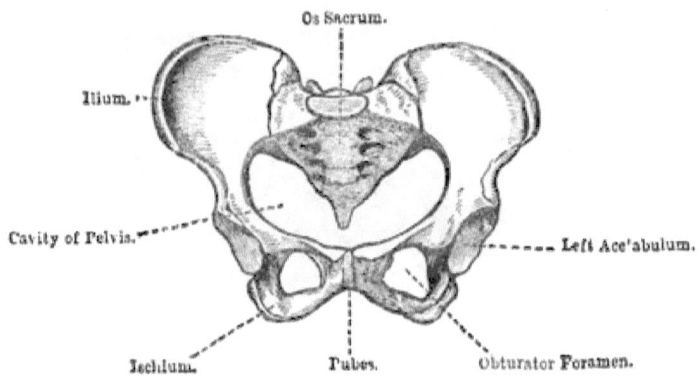

Fig. 12. The Pelvis.

to the lower limbs which are attached by the rounded
heads of the femoral bones.

It is formed by the junction of four bones—the two
ossa innominata, the os sacrum, and the os coccyx.

The *ossa innominata* are the two *hip* bones of the *pelvis*.   Each
hip bone contains a deep round *cavity*, the *acetabulum*, into which
the large rounded head of the *os femoris* (thigh bone) fits, and in
which it is retained by the *ligamentum teres* (round) the *cotyloid*
and other *ligaments*.

Each hip bone (*os innominatum*) consists of three bones—viz., the *ilium*, the *ischium*, the bone which supports us when sitting, and the *pubes*.

The *obturator foramen* is a large *hole* in the *hip* bone through which the *large blood-vessels*, and the *obturator* and *sciatic* nerves pass to the leg. It also serves to *lighten* the pelvis.

69. The Thorax is the osseo-cartilaginous conical or beehive-shaped cage, which contains and protects the principal organs of circulation and respiration.

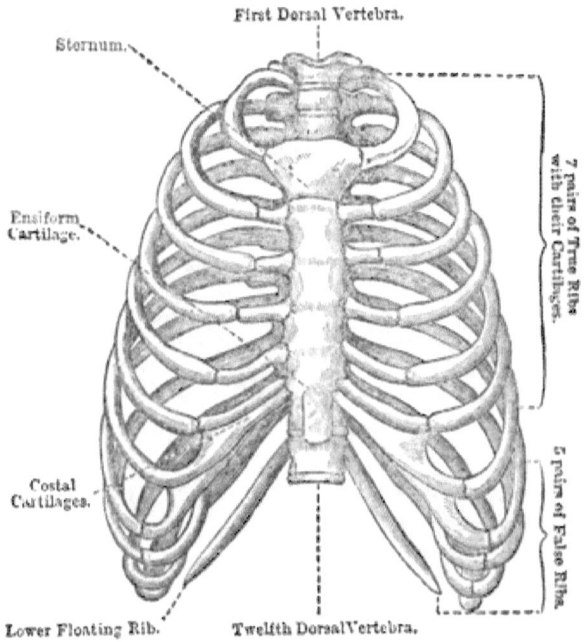

Fig. 13. The Bones of the Human Thorax.

It consists of the *sternum* (breast-bone), the lower end of which is cartilaginous (gristly)—the twelve pairs of *ribs* (more or less movable) joining the vertebra behind to the sternum in front—the twelve *dorsal vertebræ*.

70. The Costae or Ribs are elastic bony arches. They comprise seven pairs of *true* ribs and five pairs of *false* ribs.

The *true* ribs are united *directly* with the *sternum* by means of the *costal cartilages.*

The *false ribs* consist of three pairs of ribs which are joined *indirectly* to the sternum—that is, the *costal cartilages,* by which their anterior ends are terminated, join the cartilages of the last (seventh) pair of the *true* ribs, the latter only uniting with the *sternum;* and two pairs, termed the *floating ribs,* which are *not* joined to the *sternum,* and have *no* costal cartilages. (See fig. 13.)

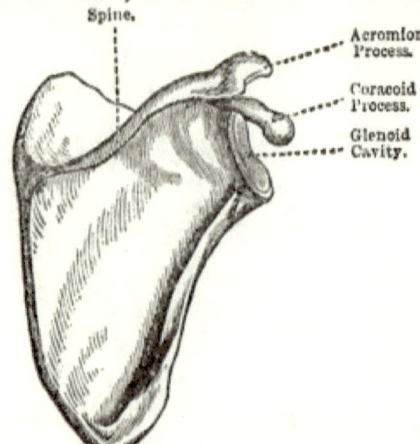

Spine.

Acromion Process.

Coracoid Process.

Glenoid Cavity.

Fig. 14. Posterior View of Right Scapula:

71. The Scapula or shoulder blade is an irregularly, somewhat triangular shaped bone. It is situated at the back of the upper side of the thorax. Its *head,* which is turned outwards, contains a *shallow* cavity or socket, the *glenoid cavity,* into which the head of the *humerus* is articulated. It is retained in its position by the *clavicle,* which is joined to its *acromion* extremity.

The *scapula* has two well marked *processes*—the *acromion process,* the summit of the shoulder, and the *coracoid process.* The general character of this bone may be well studied in the *shoulder blade* of a shoulder of mutton.

72. The Clavicle, or collar bone, is the long curved letter *ʃ*-shaped bone by which the shoulder-blade, and with it the arm, are kept in their places. (See fig. 7.)

The *clavicle* acts as a *beam,* preventing the shoulder-blade from falling forwards on to the chest, one end being attached to the top of the *sternum,* the other to the *acromion* process of the *scapula.*

73. The Arm and Hand contain (not counting the

sesamoid bones) thirty bones—viz., the *os humerus*, the *os ulna*, the *os radius*, the eight *carpal* (wrist) bones, the five *metacarpal* (palm) bones, and the fourteen *phalanges* of the fingers.

74. **The Humerus** (from Lat. shoulder), or bone of the *upper* arm, is a long, somewhat cylindrical or prismoidal hollow bone. At its upper extremity it is expanded into a globular form, the *head*, joined by a *ball-and-socket* joint to the scapula. At its lower end it is expanded into *two condyles*, at which extremity it is *articulated* (jointed) with the *ulna* and *radius* of the *lower* arm. The intermediate portion is termed the *shaft*. (See fig. 15.)

75. **The Ulna** (from Gr. *olene*, elbow) is the *larger* bone of the *fore-arm*. It is a long prismoidal bone; its upper extremity articulates by a *hinge* joint with the

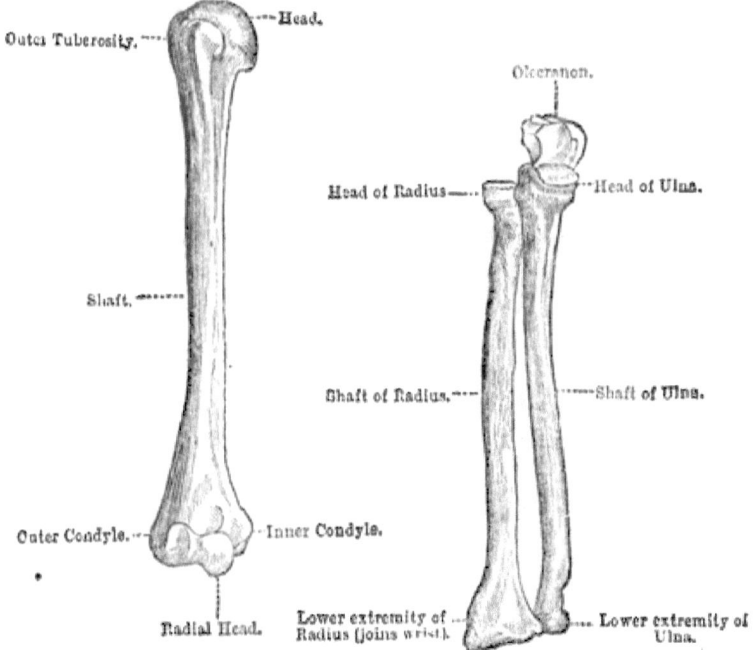

Outer Tuberosity. ---Head.

Olecranon.

Head of Radius--- ---Head of Ulna.

Shaft. --------

Shaft of Radius.--- -----Shaft of Ulna.

Outer Condyle. --- Inner Condyle.

Radial Head. Lower extremity of Radius (joins wrist). --- Lower extremity of Ulna.

Fig. 15. Humerus.   Fig. 16. Radius and Ulna.

humerus. It also articulates with the *radius.* A process termed the *olecranon* prevents the radius from bending too far back.

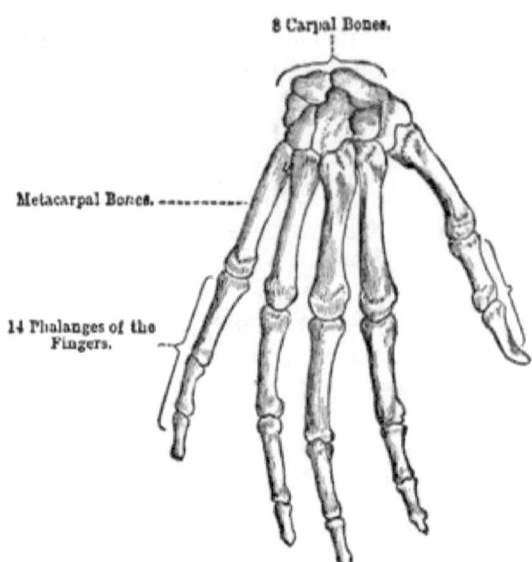

8 Carpal Bones.

Metacarpal Bones.

14 Phalanges of the Fingers.

Fig. 17. Bones of the Wrist and Hand.

76. The Radius or *spoke-bone* is the long, somewhat curved p r i s m o i d a l bone of the *fore-arm*, to which the hand is joined by the bones of the *wrist:* its upper end ar ti c u l a t e s with the *humerus.* (See figs. 16, 17.)

77. The Carpal Bones consist of the eight bones which, arranged in *two rows* of *four* each, and united by means of ligaments, form the *carpus* or wrist.

78. **The Metacarpal Bones** are the five small prismoidal bones which form the palm of the hand. They are united to the *first* row of the *phalanges* of the fingers by *hinge* joints. (See fig. 17.)

79. **The Phalanges of the Hand** are the fourteen prismoidal bones of the fingers. They are articulated, three to each finger, except the thumb, which only contains two *phalanges*, by means of *hinge* joints. (See fig. 17.)

80. **The Lower Extremities,** including the *thigh, leg,* and *foot,* comprise (excluding the *sesamoid* bones and two *patellæ*) twenty-nine bones in each *limb,* viz. :—The *os femoris* (*femur* or thigh bone) *tibia, fibula,* the seven *tarsal* bones, the five *metatarsal* bones, and the fourteen *phalanges* of the toes.

81. **The Os Femoris** (from Lat. *femur*, the thigh) is the largest and strongest bone in the skeleton.   At the top of its large globu-lar *head* is a de-pression, in which is inserted the end of the *ligamentum teres*, one of the *ligaments* by which it is retained in the *acetabulum.*

82.   The   Tibia (Lat. a flute), or *shin-bone,*originally so called from its supposed   resem-blance to an ancient *musical pipe*, is the long   prismoidal vertical bone which forms the *front* and *inner* side of the *lower* leg.   After the *os femoris*, it is the largest bone in the body.   Its head articulates   by   a *hinge joint* with the   *femur.*   Its

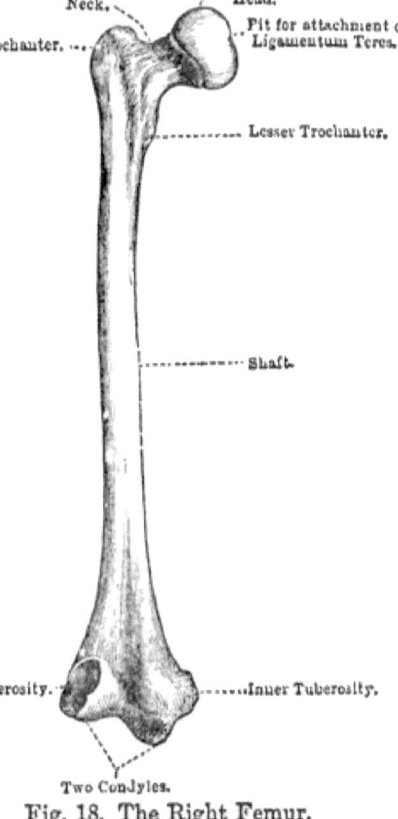

Neck.
Head.
Great Trochanter.
Fit for attachment of Ligamentum Teres.
Lesser Trochanter.
Shaft.
Outer Tuberosity.
Inner Tuberosity.
Two Condyles.

Fig. 18. The Right Femur.

lower extremity *articulates* with the *astragalus* (one of the tarsal bones).   (See fig. 19.)

83. **The Fibula** (Lat. a buckle), or *splint-bone*, is the long slender outer bone of the leg.   It is *parallel* with the *tibia*, being immovably attached to it by its upper and lower extremities, in order to increase its strength. (See fig. 19.)

84.   **The Tarsal Bones** (from Gr. *tarsos*, sole of the foot) comprise the seven irregularly shaped bones which form the heel, the ancle, and part of the sole—viz., the

os calcis, astragalus, cuboid, scaphoid, and the internal, middle, and external *cuneiform* bones.

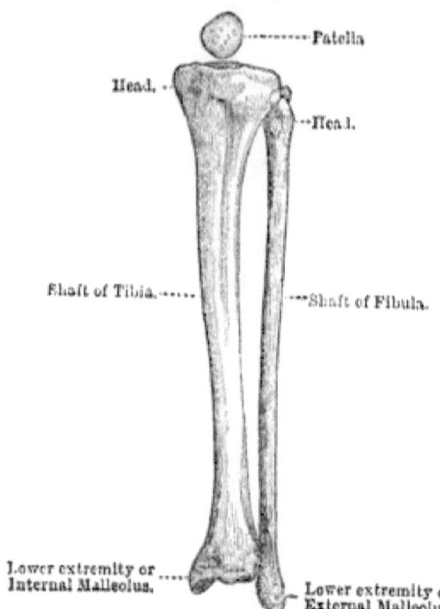

Fig. 19. The Left Tibia and Fibula.

85. The Metatarsal Bones (from Gr. *meta,* beyond, and *tarsos*) comprise the five long bones which form the *lower instep,* or front part of the arch of each foot. (See fig. 20.)

86. The Phalanges of the Foot, or the *toes,* are the fourteen bones of the toes which correspond in *number* with those of the hand. The two *phalanges* of the big toe differ, however, in articulation from those of the thumb, in not being opposable to the rest of the *foot.* (See fig. 20.)

87. The Ligaments (from Lat. *ligo,* I bind) are the flexible, very *pliant,* but *tough, inextensible,* white, shining, somewhat silvery-looking bands of *white fibrous* (connective) tissue, by which the ends of the movable bones are connected together so as to form the *movable joints:* as the ligaments of the wrist and the foot, the *transverse ligament* of the *atlas,* the *glenoid ligament* of the shoulder, the *ligamentum teres,* and the *capsular ligament* of the head of the *thigh bone,* &c.

Some few ligaments, as the *ligamenta subflava* (from Lat. *flavis,* yellow), which connect the adjacent *arches* of the vertebræ, and the *ligamentum nuchæ* of the neck of the horse, the rudiments of which only exist in man, consist almost entirely of *yellow elastic tissue.* In these cases the *elasticity* of the ligament is intended to act as a partial substitute for *muscular power*.

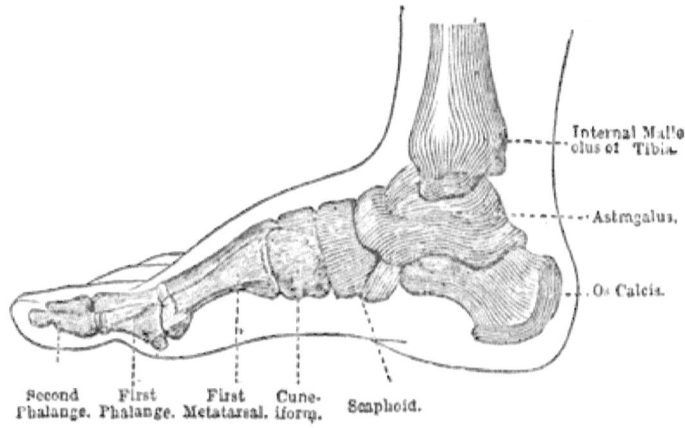

Fig. 20. Bones of the Right Foot.

For **Articular Cartilages** and **Synovial Sacs**, see *Animal Mechanics.*

CHAPTER IV.

CHEMICAL PRELIMINARIES.

88. Analysis of the Animal Body.—If the body of one of the higher animals were first *separated* into its various *complete* parts, only such parts being considered complete and distinct which, like the eyes or the stomach, performed a special and distinct office or duty, such a part would constitute an *organ.*

If such a part or organ was again submitted to disintegration, by which it was resolved into its simplest or most elementary structures, such structures would constitute the *tissues* of the organ or body.

If the *tissues* were again reduced to a simpler form, so that *all traces of structure disappeared*, the *substances* thus obtained would constitute the *proximate* or *organic principles* of which the *tissues were composed*—as *albumen, fibrin,* &c.

. If the *organic* or *proximate principles* were again in their turn *disintegrated* or *decomposed,* so that the elements entering into their composition should be ultimately reduced to their *simplest* and most *elementary* state, the substances thus obtained would constitute the *ultimate* or *chemical elements* of the *tissue* or of the body, as *oxygen, carbon, nitrogen, phosphorus, sodium,* &c.

89. A Chemical or Ultimate Element is therefore a substance which cannot be *decomposed* or *separated* into any *simpler* form of matter than itself.

Chemists are acquainted with sixty-four or more elementary substances. Of these, however, only about one-fourth have been found *as constituents* in the *animal body* in any appreciable quantity. The following are the *chemical elements* which may be obtained by the *chemical analysis* of the human body:—*Oxygen, hydrogen, carbon, nitrogen, sulphur, silicon, phosphorus, chlorine, fluorine, potassium, sodium, calcium, magnesium, iron,* and possibly *manganese,* and one or two other elements.

90. Organogens (from Gr. *organon,* an instrument; and *gennao,* I produce).—The bulk of most *animal* and *vegetable* substances is composed of the first *four* elements named in the preceding list—viz., *oxygen, hydrogen, carbon,* and *nitrogen*—these bodies have therefore been termed the *organogens.*

91. Oxygen is the most abundant element in the universe, forming *more* than *one-half* of the known crust of the earth, *eight-ninths* of all the water that exists, and upwards of *two-thirds* by weight of the human body.

Oxygen, when *free*—that is, in its pure and gaseous state—is a colourless, *tasteless, odourless,* transparent, invisible, *respirable* gas. It is itself uninflammable, but is the best *supporter of combustion* of common inflammable substances known. It is also a *supporter of life*, and is capable of entering into *chemical union* with most of the other *chemical elements*.

In appearance, and by smell or taste, *oxygen* is indistinguishable from common atmospheric air.

EXPERIMENT I.—Kindle a long splinter of wood, blow out the flame, and then plunge the *red-hot* end into a bottle of *oxygen:* it will be immediately *re-kindled*, and will burn with great vigour and *splendour*. In this way *oxygen* may be *readily* distinguished from *common air*.

EXPERIMENT II.—After the match has been burning for a few seconds, pour in a small quantity of clear transparent *lime-water:* it will immediately become *white* and *turbid*, showing that the *wood* has become *oxidized*, the *oxygen* having combined *chemically* with its *carbon*, and formed *carbonic acid gas;* another portion of the *oxygen* will also have *oxidized* its *hydrogen*, and formed water. *Oxygen gas* supports life by *oxidizing* or combining with (*burning*) the tissues. The products of this combustion or oxidation either pass off from the body in the *gaseous* form, as *carbonic acid*, the *vaporous* form, as *aqueous vapour*, or in the *liquid* form, as *solution of urea*.

92. Hydrogen (from Gr. *hudor*, water; and *gennao*, I produce) is a *chemical element*, which is never found *free* in nature. When uncombined with other elements, it is a colourless, tasteless, odorless, transparent, invisible gas. It is highly *inflammable* and *respirable*, but a *non-supporter* of *life* and *combustion*. When *burnt* (oxidized) in *air* or *oxygen gas*, it produces *watery vapour*, great *heat* being evolved. When *fatty substances* are *slowly oxidized* in the body, the *hydrogen* contributes to the animal heat. The *hydrogen* of *sugar*, *starch*, and *gum*, being already fully oxidized, cannot contribute to the *animal heat*.

EXPERIMENT.—Hold, for a moment or two only, a *cold* glass tumbler over a candle or gas flame. The sides of the tumbler immediately become covered with a deposit of *dew* (condensed aqueous vapour). The water is formed by the combination of the *oxygen* and *hydrogen*, in the proportion of *one atom* of the former to *two atoms* of the latter.

93. Water consists of the oxide of hydrogen (hydric

14 E.   D

oxide). It is, as just shown, produced whenever *hydrogen* or any of its compounds are burnt. It exists largely in nature in the form of a transparent, colourless, inodorous, tasteless, bland liquid, which *boils* at 100° Centigrade (212° Fahrenheit), and *freezes* at 0° Centigrade (32° Fahrenheit). It forms a very large proportion of the substance of all living animal bodies, and, though in itself it is *quite innutritious* (containing no carbon or nitrogen), an adequate supply of it is essential to life. The *animal heat* is kept down to its proper limit or degree by the escape of *water* in its *vaporous* condition, which thus carries off the *excess of heat*. The *chemical symbol* of one atom of water is $H_2 O$.

94. **Nitrogen** (from Gr. *nitron*, saltpetre; and *gennao*, I produce) is the most characteristic *element* of *animal* substances. It has therefore been described as the *basis* of the *animal tissues*—being to the *animal* world what *carbon* is to the *vegetable* world.

In its *uncombined* state, it exists in the form of a *permanent*, colourless, tasteless, inodorous, invisible *respirable gas*, remarkable for its series of *negative* properties. It is inflammable, and will *neither support life* nor *combustion*. Its most important ordinary compounds are *nitric acid* (nitric anhydride or pentoxide) and *ammonia*. The former consists of *nitrogen* and *oxygen* ($N_2 O_5$), the latter of *nitrogen* and *hydrogen*. It also forms the principal constituent in *bulk* or volume of the *atmosphere*, though its action is purely *negative*.

It is an essential constituent of the *proteids*, and of the so-called *flesh-forming*, *albuminous* or *plastic* food. Without *nitrogen* the *tissues* could not be repaired.

EXPERIMENT I.—Get a wide-mouthed bottle, containing *common air* only, also a stand, to support a small piece of *phosphorus*, about three inches, over a basin of water. Kindle the *phosphorus*, and immediately place the bottle over the burning phosphorus, its mouth *inverted*, and a little under the water.

The *phosphorus* at first burns brilliantly, but is speedily extinguished; the water now quickly rises inside to about one-fifth the height of the bottle. Allow the white fumes to subside: the clear transparent gas which now fills the bottle is *nitrogen*.

EXPERIMENT II.—Plunge a *lighted* match or candle into the bottle, it is immediately *extinguished*, showing that it does not support combustion.

The last experiment shows that nitrogen gas is a mere *diluent* of the *oxygen* of the atmosphere; or, in other words, that it plays the same part in *diluting* and lessening the energy of the *oxygen* that the water plays in diluting the spirits in a glass of toddy.

95. Ammonia (spirits of hartshorn) is almost invariably obtained by the *decomposition* of *animal* or *vegetable* matter, containing *nitrogen* either *slowly*, at ordinary temperatures, or *quickly*, by the application of *heat*.

In the form of gas it is colourless, transparent, and *invisible;* has a very strong, peculiar, *pungent* odour ; is very irritating, and is *irrespirable*. It is very *soluble* in water, forming with it a strong *aqueous solution*, having all the *chemical* properties of the gas. This is usually sold as *ammonia*.

It is strongly *alkaline* turning *red*, betimes *blue*. It exists in the atmosphere in *minute* quantities. It is from the *ammonia* in the *air* that *plants* obtain their *nitrogen*.

One *molecule* of ammonia consists of 1 *atom* of *nitrogen*, or 14 parts by weight, chemically united with 3 *atoms*, or 3 parts by weight of *hydrogen*. Its chemical symbol is $N\,H_3$.

The *urea* which leaves the body, carrying with it the *waste nitrogen* of the *tissues*, afterwards splits up into *carbonic acid* and *ammonia*.

96. Atmospheric Air (See **Respiration**).

97. Carbon (from Lat. *carbo*, coal), is usually regarded as the chemical *basis* of the *vegetable* world. It is one of the principal constituents of most vegetable and animal substances.

Good *charcoal* consists almost exclusively of *carbon;* but it exists in its purest known form in the *diamond*.

Carbon in its *free* state is a solid, infusible, fixed, insoluble, *combustible* substance, having a strong attraction for oxygen, with which it readily combines at a high temperature, producing *carbonic acid gas* (carbon dioxide).

It is an essential constituent of all *food stuffs*, especially of *heat-forming, fuel*, or *respiratory* food. (See Food.)

98. **Carbonic Acid Gas** (carbonic dioxide) is the *noxious*, and, when *pure* (concentrated), *irrespirable* gas which is given off from *lime-kilns*, effervescing soda-water, ginger-beer, champagne, &c.; or when strong vinegar is poured on to chalk, or on to egg-shells; and which is produced when wood, coal-gas, and ordinary inflammable substances are burnt.

It is also produced during respiration, and forms $2\frac{1}{2}$ to 5 per cent. of the air expelled from the lungs.

It is a *heavy*, transparent, colourless, *uninflammable* gas, and is also a *non-supporter* of *life* and *combustion*, extinguishing flame, and, when inhaled, quickly producing death from *suffocation*. It has the properties of an *acid*, turning solution of *blue* litmus *red*. It is *soluble* in water and in the blood.

Carbonic acid is a very *heavy* gas, and therefore tends to collect at the bottom of old wells, caves, beer vats, &c., sometimes producing fatal results, when men enter them carelessly before ventilation.

Each *molecule* of carbonic acid contains 1 *atom*, or 12 parts by weight of carbon, *chemically united* with 2 *atoms*, or 32 parts by weight of oxygen. Its chemical symbol is CO_2.

EXPERIMENT I.—Introduce a small quantity of powdered carbonate of soda, or powdered chalk, or "whitening," into a large wide-mouthed bottle, and pour some strong vinegar into it. The powder will immediately begin to *effervesce*. After a few moments, introduce a lighted candle or match into the bottle, it will immediately be *extinguished* by the carbonic acid evolved.

EXPERIMENT II.—Hold the mouth of the bottle containing the effervescing mixture over the mouth of a glass tumbler. Plunge a lighted match or taper into the tumbler: the flame of the candle or match will be immediately extinguished, as before,—thus showing that *carbonic acid* is much heavier than common air; and consequently, though gaseous, may be *poured* from one vessel into another.

99. Putrefaction, **Decomposition,** and **Decay** after death, consist of the series of changes which ensues in most *complex organic* (especially *nitrogenous*) substances, under the combined action of *water* and *oxygen*, by which they first split up into *simpler* forms or compounds, and then become more or less oxidized. The *offensive odour* evolved from bodies passing through this state of *rottenness* is chiefly due to the presence of *carbon, sulphur,* and *phosphorus,* the larger portions of which are eliminated in the form of *carburetted, sulphuretted,* and *phosphuretted, hydrogen* gases.

100. *Nitrogen* is distinguished by its *feeble* power of *chemical attraction* for the other elements in its compounds; therefore the latter tend speedily to *break away* from it. *Oxygen,* on the other hand, characterized by its *powerful* attraction for these elements, promotes this process of *splitting-up* by, as it were, *chemically pulling* them *away* from the nitrogen *to* itself.

101. **Incidental Elements.**—In addition to the *organogens* (sec. 90), which are *essential* elements of the animal body, a number of other chemical elements, as previously stated (sec. 89), are usually present. These are described as the *Incidental Elements,* among the more important of which are sulphur, phosphorus, chlorine, sodium, calcium, and magnesium, the latter of which form the *bases* of the earthy salts so largely present in the body.

102. **Mineral Compounds.**—The principal mineral compounds of the body are sodium chloride (common salt), and a calcic phosphate (bone phosphate of lime), of which there are 5 or 6 lbs. in the body, calcium carbonate, and sulphate (carbonate and sulphate of lime), and the alkaline carbonates and phosphates.

After much *mental* exertion or *nervous* exhaustion, the quantity of the *phosphates* excreted in the urine as *acid* phosphates, increases very greatly as the result of nervous tissue *waste.*

It may be as well to state, for the benefit of the non-chemical reader, that the *salts* here mentioned are simply *compounds* of sodium, lime, magnesia, &c., with chlorine, sulphuric, phosphoric, carbonic, or other acids.

103. Endosmosis, Exosmosis, Osmosis (from Gr. *osmos*, impulse) **is a** species of *physico-chemical* action which prevails largely in living bodies, more especially in the various processes of *nutrition, secretion,* and *respiration.*

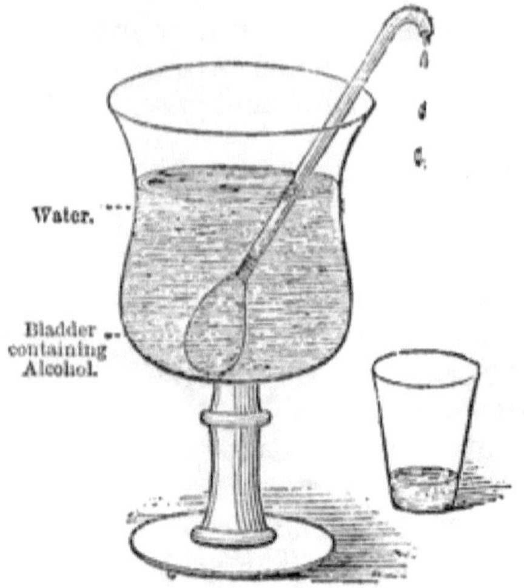

Osmosis · is the action or process by which two different liquids separated from each other by a *porous solid,* or an intervening *membrane* will pass through its substance and *mix* together on either side of it. Usually a very much *larger* proportion of the one fluid than of the other *passes* through the membrane.

Water.

Bladder containing Alcohol.

Fig. 21. Illustration of Osmosis.

EXPERIMENT.—Procure a *glass vessel,* a *glass tube* with a small bore and a curved end, a *small bladder,* some *water,* and some *alcohol* (spirits of wine). Fill the vessel two-thirds full of water, tie the bladder to the lower end of the tube, fill it with alcohol, and immerse it in the water. (See fig. 21.) The water will *pass* in through the walls of the bladder by *endosmosis* until the tube not only becomes full but overflows in drops, which may be collected and measured, in which case the instrument becomes an *endosmeter.* A very small quantity of the alcohol *passes out* of the bladder by *exosmosis.*

If a thin *collodion* bag were substituted for the bladder, the process would be reversed, and much *more* alcohol would pass *out* of the collodion bag than water would pass *in.*

The two Greek prefixes *endo*, in, and *exo*, out, show the *direction* in which the liquid passes.

Mixed solutions may be analyzed on this *principle;* the process is then termed *dialysis.*

104. Diffusion of Liquids and Gases.—When two *liquids* or *gases* of unequal densities are mixed, they will interpenetrate each other's substance, the *lighter* gas or liquid proceeding downwards, the *heavier* gas or liquid upwards *against gravity* until they are uniformly mixed. In this way more heavy carbonic acid may be found on the tops of the highest mountains than on the level of the plains below.

It is by this principle of *diffusion*, which is governed by definite laws, that the fresh air taken in during each inspiration mixes with the air in the lung sacs, the *stationary air*, in the language of Professor Huxley, acting "the part of a middleman between the two parties—the blood and the fresh *tidal air.*"

105. A Proximate Principle is an organic compound which enters into the substance of a tissue or organ. Its *molecules* are exceedingly complex, usually containing the four *organogens* united in very complex proportions, together with small quantities of sulphur and phosphorus.

Proximate principles comprise two kinds—the *nitrogenous*, termed the *proteids*, and the *non-nitrogenous.*

The principal *nitrogenous* proximate principles which enter into the composition of the animal body or its secretions are albumen, fibrin, syntonin, casein, globulin, hæmatine, gelatine and chondrin, and keratin.

106. Albumen (from Lat. *albus*, white), is the chief *nitrogenous* constituent of the blood and the "white of egg." It derives its name from its opaque *white* appearance when boiled. It is soluble in *alkaline* solutions. In its ordinary state, as in the *serum* of the blood, it *coagulates* when acted upon by *heat* or *acids.*

107. Fibrin is a *nitrogenous* substance which closely resembles *albumen* in its chemical composition and properties; it differs from it, however, in being *spon-*

taneously coagulable. It forms the net-work in blood-clot. It derives its name from its spontaneous tendency to form fibres.

108. **Syntonin** is the variety of *fibrin* constituting the bulk of *muscular fibre.*

109. **Casein** (from Lat. *caseus,* cheese) is the characteristic, most valuable and *nutritious* constituent of *milk.* It is separated as curd, on the addition of acid, or, when milk turns *sour,* from the conversion of its *sugar* into *lactic acid.* Casein is chemically identical with the *legumen* of beans, peas, and lentils.

110. **Globulin** is simply that variety of *albumen* of which the red corpuscles of the blood are chiefly composed. It exists in the crystalline lens of the eye as *crystallin.*

111. **Gelatin** (from Lat. *gelu,* ice), though possessing the general properties and chemical composition of the *proteids* or *albuminoids,* probably does not exist in the body until it has been developed by the prolonged action of boiling water on fibro-cartilage, the skin and other substances containing *white fibrous tissue.*

Isinglass, size, and glue are forms of gelatine. Its solutions possess a remarkable power of solidifying on cooling. A hot aqueous solution, containing only 1 part of gelatine to 99 parts of water, will solidify into a jelly on cooling.

112. **Chondrin** (from Gr. *chondros,* cartilage), in general resembles *gelatin,* but is obtained by the action of hot water on *true cartilage.*

113. **Keratin** (from Gr. *keras,* horn), is the peculiar principle of the horny tissues, including horn, hair, hoofs, and whalebone.

114. **Protoplasm** or **Bioplasm** is the general term applied to the supposed *nitrogenous albuminoid* or *proteid* substances or bases, or formative matters, out of which the various tissues are built up. It is supposed to be present in the bodies of all *living* animals and *growing* tissues, and has been variously termed sarcode, blastema,

and germinal matter. It consists of carbon, hydrogen, nitrogen, and oxygen, in about the same proportions as in "white of egg" (albumen). All forms of *protoplasm contract* under the stimulus of the *electric current*, and "stiffen" (coagulate) under the influence of *heat*.

115. Non-Nitrogenous Principles.—The principal non-nitrogenous substances used by animals as food are the *amyloids* and the *fats*. They consist of oxygen, hydrogen, and carbon.

116. The Amyloids (from Gr. *amulon*, starch) comprise those bodies in which the *oxygen* and the *hydrogen* are already combined in the proportion in which they form *water*, and from which, therefore, no further heat can be derived by the body. They comprise *starch, gum, dextrin,* and *sugar*, and are only useful as *heat-formers* because of the *carbon* they contain.

117. Starch is an insoluble vegetable substance, and therefore, before it can be utilized as food, must be converted in the alimentary canal into soluble *sugar*.

Its chemical symbol is $C_6 H_{10} O_5$, a molecule of starch thus consists of carbon, hydrogen, and oxygen in the proportion of 6 atoms of *carbon* and 5 atoms of *water* ($H_2 O$).

118. Dextrine and **Gum** contain the same chemical elements as starch, united in the same proportions, but *grouped* differently; they differ from starch mainly in their great *solubility*. It is, however, uncertain how far they are useful as food, especially the latter.

119. Sugar is mainly of vegetable origin; it is distinguished by its sweet taste, solubility, and crystalizability, and its tendency, under favourable circumstances, to undergo *vinous fermentation*, in which *alcohol* is produced and carbonic acid evolved.

Its chemical symbol is $C_{12} H_{22} O_{11}$, that is, one molecule of sugar comprises the elements of 12 atoms of *carbon* and 11 atoms of *water*, therefore the *carbon* only is useful as *fuel food*.

When absorbed into the blood its carbon is either burnt as *respiratory* or fuel food, or the elements of the sugar are converted into fat.

120. The Fats are oily, *non*-nitrogenous substances, consisting of carbon, hydrogen, and oxygen, in varying proportions, which are chiefly found in the bodies of animals. They contain a large excess of *carbon* and *hydrogen*, as compared with oxygen, in consequence of which their *heating* powers (as fuel *food*) are very great.

They are insoluble in, and will not mix with water, but are converted into soluble *soaps* by the *alkalies*. They are either liquid or readily fusible, and are soluble in *ether* and *hot alcohol.*

121. Organ—Organized Bodies (from Gr. *organon*, an instrument). An *organ* is a special or distinct part of the body, which performs a special action, function, or office, as the eye, the ear, the kidney.

A body consisting of a number of organs united into *one system*, and *acting* together for a *common* object, is termed an *organized body*, or an *organization*. If such an organization contains a very *few* organs only, or if it consists of a great multiplication of the same organs, it is said to be of *low* organization ; if it consists of a great *many* organs, each of which performs a *distinct* function, it is said to be of *high* organization.

Organized bodies differ from *inorganic* or *mineral* bodies, chiefly in the greater *complexity* of their chemical composition, their complex, heterogeneous, and cellular or vascular structure, and their growth, both by *interstitial addition*, and *external* deposit.

122. The Function of an organ is the action, use, office, or duty performed by it, as sight, hearing, and *excretion.* The functions of digestion, absorption, respiration, nutrition, secretion, excretion, circulation, reproduction, &c., *common* to both animal and vegetable life, are described as *vegetative* functions; those peculiar to animal life only, as spontaneous locomotion, sensation, thought, are described as *animal* functions, or functions of *relation.*

123. Biology (from Gr. *bios*, life, and *logos*, a discourse), or the science of life, comprises two leading divisions, botany and zoology, the former, in its larger sense, treating of all that belongs to *plant* life, the latter of *animal* life.

124. Anatomy (from Gr. *ana*, through, and *temno*, I cut), is the science which treats of the *form, position,* and *structure* of the various parts of *organized* bodies. It is studied mainly by means of *dissection*, or, though less perfectly, by diagrams *drawn* or *photographed* from *dissections*.

125. Physiology and **Pathology** (from Gr. *phusis*, nature, and *logos*, a discourse) is the branch of biological science which treats of the uses and the modes in which the various functions of the body are performed during health; it is therefore sometimes defined as the *science of health*. The science which treats of the modes in which the *organs* perform their *functions* during *disease* is termed *pathology*.

126. Histology (from Gr. *istos*, a web or tissue, and *logos*, a discourse) is the branch of biological science which treats of the exceedingly *minute* or microscopical structures of the *tissues* and their functions.

CHAPTER V.

HISTOLOGICAL PRELIMINARIES.

127. The Epithelium is probably one of the *simplest* structures in the body. It consists of one or more *layers* of microscopic *nucleated* cells, termed *epithelial cells*, which are arranged so as to form *membranes*, which line (and are found only on) the *free surfaces* on the *interior* and on the *exterior* of the body, thus forming the exterior or free surface of the *epidermis*, and of the *mucous* and *serous* membranes.

Epithelial cells consist of an outer exquisitely fine *cell-wall*, a *nucleus* and *nucleoli*, and sometimes also other cell contents of fluid or granular matter. They are

classified under *four* principal forms or varieties, according to their *shape* or other characters, viz. : 1, *Squamous*
or *tesselated;* 2, *Spheroidal* or *glandular;* 3, *Columnar* or
cylindrical; and 4, *Ciliated.*

128. **The Squamous or Tesselated Epithelium Cells**
consists of flattish cells, which overlie each other like

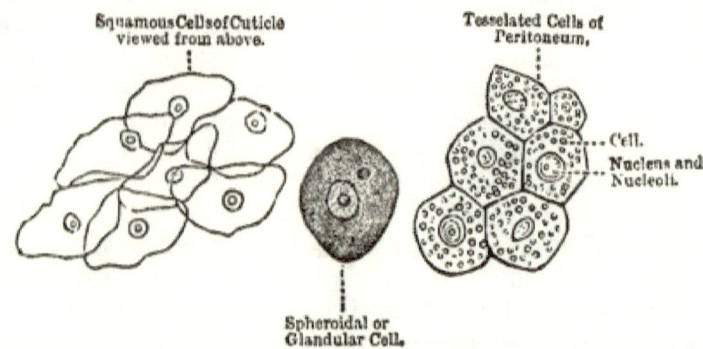

Fig. 22. Squamous and Spheroidal Epithelium Cells.

the *scales* of a fish, as in the cuticle, or which are placed
side by side, edge to edge, like tiles or stones in the
pavement, as in the *serous* and *synovial membranes,* the
interior of the *lymphatics* and *blood-vessels.*

These cells are sometimes charged with *pigment,* as
in the choroid coat of the eye; they are then termed
pigment cells.

129. **The Spheroidal or Glandular Epithelium Cells**
consist of the *rounded* or *globular* cells which line the
interior of the compound glands, as those of the *liver,
gastric glands,* &c.   They assume a *polygonal* shape
under pressure when crowded together.

The *glandular* epithelial cells really do the *secretory*
work of the gland.

130. **The Columnar or Cylindrical Epithelium Cells**
consist of the more or less oblong *cylindrical,* or conical-

shaped shells, which, placed side by side, standing perpendicularly on their *lower* or *attached* extremities
(which are in general *smaller* than their *free* ends), line
the surfaces of the stomach and intestines, including

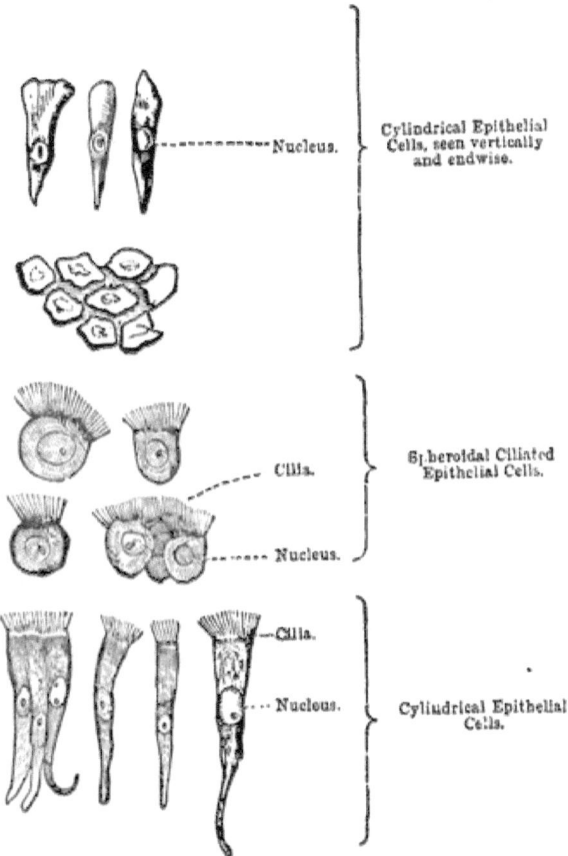

Fig. 23. Cylindrical and Ciliated Epithelial Cells.

the walls of the villi and of the *upper* portions of the
gastric follicles, and also of the gland bladder, &c.

131. **The Ciliated Epithelium Cells** consist of cells in
general of the *cylindrical* variety, the *free* and *expanded*
extremities of which are covered with exquisitely fine,
pliant, *microscopic vibratile processes*, termed *cilia*.

They line the *free* surface of the entire respiratory tract, including all the air passages and tubes down to the air cells. A *tesselated* variety of *ciliated epithelium* also lines the *ventricles* of the brain, and the *central canal* of the spinal cord.

132. **Connective Cellular or Areolar Tissue** consists of a mesh-work in which large quantities of *white fibrous tissue* are intermingled with a comparatively small quantity of *yellow elastic tissue.* This tissue is most abundantly distributed through the body, forming a sort of *matrix*, which interpenetrates and *invests* the various organs, and *binds* their tissues and structures together.

It is a very pliant, flexible, elastic, extensible, whitish-looking structure. It probably itself receives no nerves and but very few blood-vessels.

Fig. 24. White Fibrous Tissue.

Showing larger and smaller wavy bundles of parallel filaments. Magnified 40 diameters.

133. **White Fibrous Tissue** consists of bands of parallel wavy fibres or filaments, $\frac{1}{10000}$ to $\frac{1}{20000}$ of an inch in diameter. It is exceedingly tough and flexible, but *inextensible* and inelastic. It contains but few nerves and blood-vessels. It forms the chief constituent of, 1, Connective tissue; 2, Ligaments; 3, Tendons; 4, Fibrous membranes, as the *periosteum, dura mater*, and the *sclerotic coat* of the eye. It yields *gelatine* on boiling

When a filament of connective tissue is treated with *acetic acid* its white fibrous tissue swells up enormously, entirely losing all character of fibre, the yellow elastic fibres being visible under the microscope as fine sharp lines in the middle of the swollen mass.

134. Yellow Fibrous Tissue consists of exceedingly fine, sharp, well-defined, microscopic, cylindrical, flexible, extensible, elastic, fibres, about $\frac{1}{40000}$ of an inch in diameter. It is more or less sparingly distributed through *connective* tissue, but it forms the bulk of certain *elastic* structures, as the *ligamenta subflava* and the *vocal cords*. It is nearly as *elastic* as *india rubber*. It does *not* yield *gelatine* when boiled.

A variety of *yellow elastic tissue*, the filaments of which *anastomose* very freely with each other, and which forms the bulk of the *middle* coat of the larger arteries, is known as *fenestrated membrane*.

When torn, its ends *curl*

Fig. 25. Yellow Elastic Tissue.

Fibres from the Ligamenta Subflava, some of which branch off into smaller curling fibres, and some of which anastomose with each other. Magnified 200 diameters.

up, thus frequently, as when limbs are torn off by machinery, retracting into, so as to plug up and stop the torn ends of the arteries, so that little or no blood is lost from them.

135. Adipose Tissue simply consists of *fat cells* distributed through the *meshes* of the *connective tissue*. The fat cells, about $\frac{1}{500}$ or $\frac{1}{400}$ of an inch in diameter, consist of oval or globular cell-walls, formed of a fine, transparent, and structureless membrane, filled with a yellowish oily fluid. After death, when the animal temperature falls, this oily fluid solidifies or coagulates, and becomes hard, as in the case of mutton suet. *Adipose tissue* is more or less *vascular*, the cells being more or less held together by the capillaries.

Its chief uses are :—(1.) It serves as a store of *heat-forming* material, which may be re-absorbed into the blood and burnt when required. (2.) As a *bad conductor* of *heat*, it tends to prevent its escape from the surface of the body. *(3.)* It serves as *packing material*, filling up space, forming a bed for and protecting the softer organs.

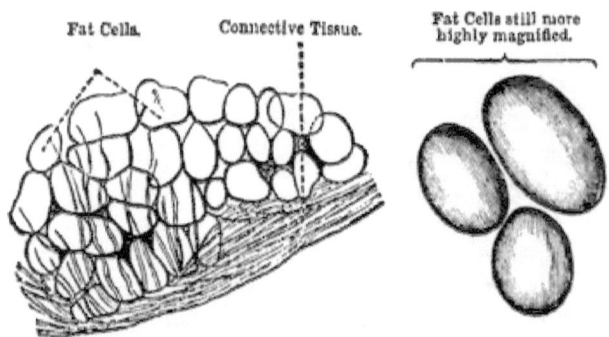

Fat Cells.　　　Connective Tissue.　　　Fat Cells still more highly magnified.

Fig. 26. Adipose Tissue (Fat Cells).

136. Cartilage or Gristle.—In the very young state of

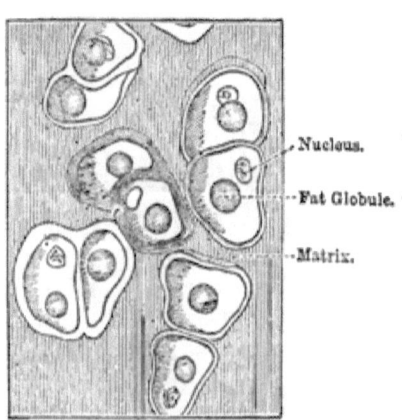

Nucleus.

Fat Globule.

Matrix.

Enclosed in Cartilage Cell.

the child the *skeleton* consists entirely of *cartilage,* which disappears as *ossification* takes place; this is therefore termed *temporary* cartilage.

Permanent cartilage comprises two varieties—(1),*Hyaline* or *articular ;* (2), Fibro - cartilage.

Articular, hyaline, or *true cartilage,* that by which

Fig. 27. Section of Articular Cartilage.

Magnified 350 diameter

the *ends* of the bones forming the *movable* joints are *tipped,* is a firm, flexible, extensible, tough, elastic, whitish, opalescent substance. It consists of a *matrix* having somewhat the appearance of ground glass, in which are imbedded a large number of irregularly shaped *nucleated* cells, $\frac{1}{1200}$ to $\frac{1}{900}$ of an inch in diameter. It is usually described as *non-vascular*, and is not supplied with nerves. (See fig. 27.)

The *medullary cavities* in the long bones are formed as follows:— (1.) *Bony tissue* is formed or deposited *around* the sides, and at the ends of the cartilages, which form the skeleton of the very young subjects. (2.) The bony matter goes on increasing by *external* addition, while the original internal cartilage becomes *absorbed*.

137. Osseous Tissue.—Bony tissue is of two kinds— *cancellous* and compact.

Cancellous bony tissue consists of a *network* of slender fibres, minute bars, or *lamellæ* of bone joined together so as to present somewhat the appearance of *lattice-work*, from which it derives its name. It constitutes the mass of the irregularly shaped bones and the *enlarged ends* of the long bones. The interstices in cancellated tissue are filled with a kind of marrow.

Compact bony tissue, which forms a thin shell on the exterior of the irregular bones and which forms the *shafts* of the *long* bones, con- sists essentially of a *series* of con- *centric* plates, or *laminæ* of bone, arranged round central *canals*, termed *Haver- sian canals*, each series forming, in fact, what may be termed a Ha- versian system.

A *Haversian* system consists,

14 E.

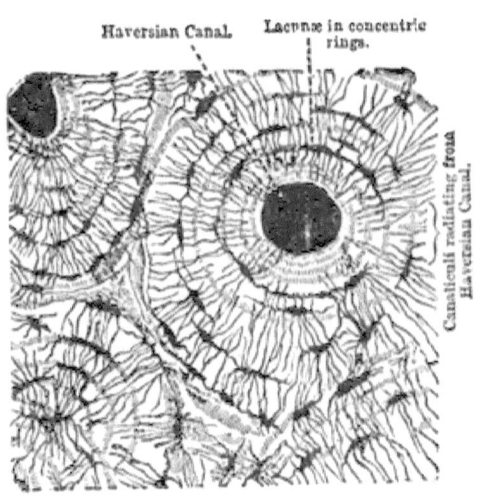

Fig. 28. Transverse Section of Compact Osseous Tissue (Bone.)

E

as shown in the diagram, of 1, A central *Haversian canal;* 2, A series of concentric bony *lamellæ;* 3, A series of concentric *rows of lacunæ,* by which the bony *lamellæ* are separated into distinct series; 4, A large number of *canaliculi, radiating* from the central *Haversian canal,* and joining the various surrounding (concentric) lamellæ into one system.

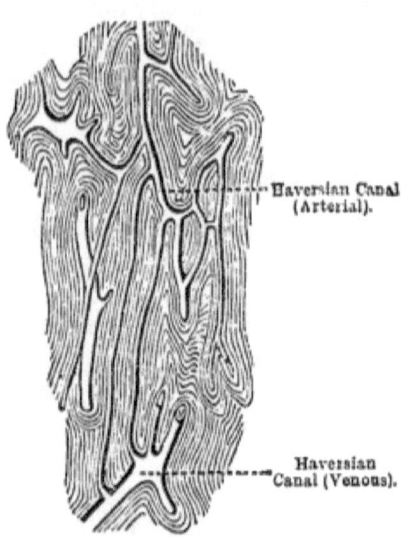

Haversian Canal (Arterial).

Haversian Canal (Venous).

Fig. 29. Longitudinal Section of Compact Bone.

The *Haversian canals* have an average diameter of about the $\frac{1}{500}$ of an inch; the longer ones contain the minute blood-vessels which convey *nutriment* to the *interior* of the bone.

The *lacunæ* are minute pits or cavities of a very irregular shape which contain nuclei. They were formerly termed the bone corpuscles, and are described by Dr. Beale as containing minute masses of *protoplasm,* or *germinal matter,* which possibly contributes to the nourishment of the bone. They are described by some physiologists as consisting of minute cavities, or gaps, formerly occupied by the cartilage cells, the whole of the surrounding cartilage having been invaded by the earthy salts during *ossification,* except the immediate neighbourhood of the *nuclei* of the cells. Their irregular outline gives them a peculiar straggling spider-like form.

The *canaliculi* are exceedingly minute canals or *tubes* which pass off and through the various *lacunæ,* appearing to *radiate* from the *Haversian canal* and connect it with the various *lamellæ* which surround it. They doubtless distribute the *nutriment* contained in the *liquor sanguinis* through the bone. (See fig. 28.)

138. The Enamel, which forms the *surface* of the

crown or exposed parts of the teeth, is the hardest, most compact, and most mineral or earthy tissue in the body; it contains about 98 per cent. of *earthy*, and only about 2 per cent. of *animal* matter.

It consists of minute, striated, hexagonal rods, *prisms*,

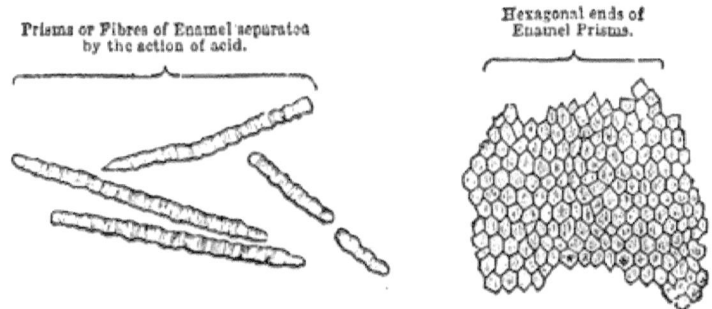

Fig. 30. Fibres or Prisms of Enamel.

or fibres, which stand endwise, side by side, perpendicularly to the surface of the tooth, or to the *dentine*. (See fig. 30.)

139. **Dentine or Tooth Tissue** constitutes the mass of

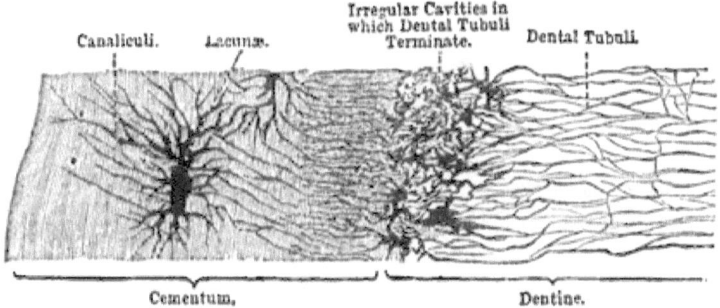

Fig. 31. Transverse Section of Tooth at the Junction of Cementum and Dentine.

the tooth. It consists of a modification of *osseous tissue*, containing, however, a much larger proportion of earthy matter (contains about 78 per cent.) than true bone.

When examined under the microscope, it is seen to consist of a dense homogeneous substance (*intertubular tissue*), which is permeated by an immense number of very minute wavy tubes (the *dental tubuli*), which anastomose with each other.   (See fig. 31.)

140. **The Crusta Petrosa**—*cementum* or *cortical substance*—is the layer or crust of true bone, which surrounds or covers the hidden portion of the tooth from the *neck* to the end of the *fang*.   (See fig. 31.)

141. **Muscular Fibre.**—The peculiar property of *muscular fibre* is its *contractility*, or power of *shortening*, under the influence of the will or of nervous stimulus, or under that of chemical, mechanical, or electrical irritation.   There are two kinds of muscular fibre—*smooth* and *striated*.   (See *Animal Mechanics*.)

142. **Non-Striated (Organic) Muscular Fibre**, also termed *smooth unstriped* muscular fibre, forms the chief constituent of the *involuntary* and of the *hollow* muscles, as those of the alimentary canal, the bladder, the gallbladder, the coats of the arteries and of the excretory ducts and larger lymphatics.   It is also found in the trachea, the iris, the skin, and elsewhere.   Its contrac-

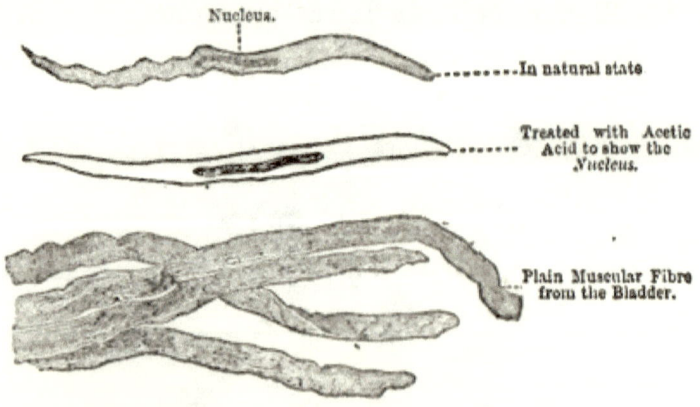

Fig. 32. Smooth (Involuntary) Muscular Fibre.

tion in the hair-sacs, from cold or fright, causes the *hair* to *rise*—thus making the hair "stand on end;" it also produces "goose-skin."

Organic muscular fibre consists of minute elongated fusiform (spindle-shaped), flattish, *nucleated, contractile* fibre-*cells* of a pale yellowish colour, about $\frac{1}{4500}$ to $\frac{1}{3500}$ of an inch in *diameter*, and $\frac{1}{600}$ to $\frac{1}{300}$ of an inch in *length*. By their union, they form minute ribbon-like filaments or fibres, which do *not* contain any *sheath* or sarcolemma. The primitive *nucleated* cells of which they are composed readily separate when treated with nitric acid.

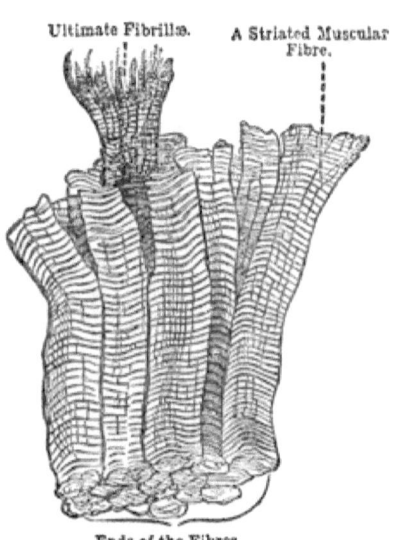

Ultimate Fibrillæ.   A Striated Muscular Fibre.

Ends of the Fibres.

Fig. 33. Portion of a Voluntary Muscle.

143. A **Voluntary Muscle**; that is, a muscle which acts according to, or is controlled by, the *impulses* of the will—consists of a *bundle of bundles* of striated muscular fibre. The smaller bundles are termed *fasciculi*, and the *sheath* of connective tissue, by which they are enclosed or invested, is termed the *fascia* of the muscle. It is abundantly supplied with nerves and blood-vessels.

The following Table shows the plan of structure of a voluntary muscle :—

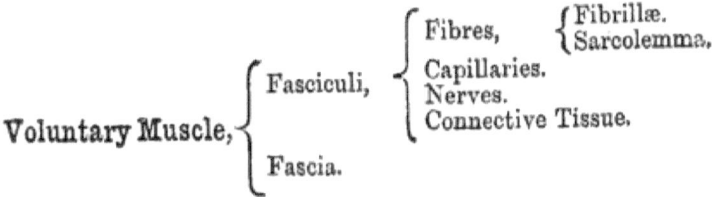

144. **Striated (Voluntary) Muscular Fibre**, as seen under the microscope, consists of minute, pale yellowish,

cross-marked, *contractile* fibres.   Each *primitive* fibre is
invested with a delicate *sheath* of fine, tough, elastic,

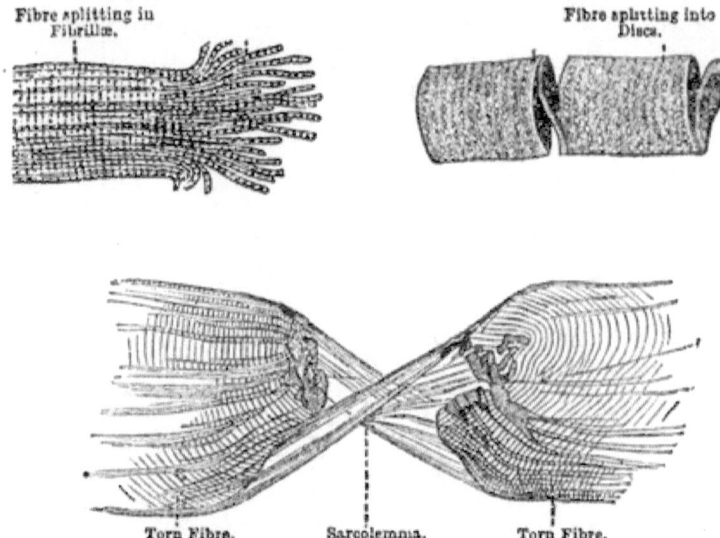

Fibre splitting in Fibrillæ.

Fibre splitting into Discs.

Torn Fibre.   Sarcolemma.   Torn Fibre.

Fig. 34. Striated (Voluntary) Muscular Tissue.
The two upper figures *devoid* of Sarcolemma.

transparent, structureless membrane, termed the *sarcolemma* or *myolemma*.

Each primitive fibre may be split up *longitudinally*
into *fibrillæ*, and *transversely* into minute discs, as
shown in the diagram.   (See figs. 33 and 34.)

The primitive fibre itself contains neither blood-
vessels nor connective tissue, but occasionally nuclei
may be seen within its substance.   The *sarcous* element,
contained within the *sarcolemma*, consists of *syntonin*.
The peculiar effect of the fine *striæ* (cross-marking) is
apparently due to the fibres being composed of a series
of *alternations* of a partially *opaque* with a more *transparent* substance.

A few hours after death the muscles of the body become hard
and rigid—undergo "death-stiffening"—or *rigor mortis*, the
various parts of the body retaining the position which they
held when the *stiffening* commenced.   This *death-stiffening* most

probably results from the *coagulation* of a *liquid* contained in the substance of the muscular fibre. The *liquid*, if squeezed out of the fibre, will *coagulate* spontaneously; after a time it again liquifies, the body becoming soft and flaccid. If the *rigor mortis* sets in soon after death, it usually lasts but a short time; if it sets in late, it usually lasts much longer.

145. Nervous Tissue, or Neurine, comprises *two* essentially distinct kinds of structure, viz. :—the *fibrous*, and the *ganglionic vesicular* or *cellular*. The former are the essential components of the *nerves*, and the *interior* of the *brain*, the latter of the *ganglia* and the *outer* layer of the *brain*, and the *inner* portion of the *spinal cord*.

146. Primitive Nerve Fibre.—Ordinary nerve fibres or tubules consist during life of soft, flexible, fragile, *transparent*, oily-looking, parallel, sub-cylindrical fibres, described as having somewhat the appearance of fine

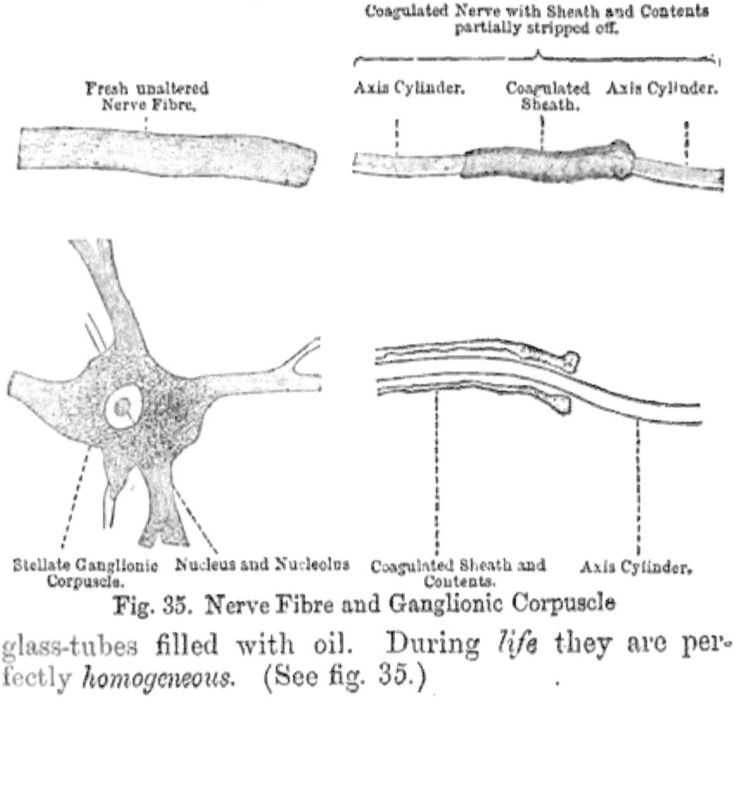

Coagulated Nerve with Sheath and Contents
partially stripped off.

Fresh unaltered Nerve Fibre.

Axis Cylinder. Coagulated Axis Cylinder. Sheath.

Stellate Ganglionic Nucleus and Nucleolus Coagulated Sheath and Axis Cylinder.
Corpuscle. Contents.

Fig. 35. Nerve Fibre and Ganglionic Corpuscle

glass-tubes filled with oil. During *life* they are perfectly *homogeneous*. (See fig. 35.)

Immediately *after death, coagulation* of the nerve substance sets in, by which it separates or *differentiates*
itself into different layers, viz.:—1, An *outer* structureless *membrane* forming a tube ; 2, An *inner* grayish,
solid axis-cylinder, which passes up the *middle* of the
tube ; 3, A fluid substance in the *interspace* between
the *axis-cylinder* and the *outer* tube.

These various structures are respectively illustrated
in the diagram of nerve fibre. (See fig. 35.)

The *axis-cylinder* of a nerve fibre is rendered very distinct, in
the beautiful microscopic slides sold by the opticians, by immer-
sion in ammoniacal solution of carmine, the *axis-cylinder* becoming
deeply-coloured red by it, while the *tube-sheath* is comparatively
unaffected, and appears as a pale ring surrounding it.

The diameter of the nerve fibres varies greatly, being
greater at their commencement, and decreasing towards
their terminations after they have left the *nerve-trunks*.
Those in the nerve-trunks are from $\frac{1}{3000}$ to the $\frac{1}{2000}$ of
an inch in diameter ; those in the *gray matter* of the
brain and *spinal cord* are $\frac{1}{15000}$ to $\frac{1}{10000}$ of an inch in
diameter.

147. **A Nerve** or **Nerve Trunk** consists of a *bundle* of
nerve-fibres surrounded by a *sheath* of connective tissue
termed the *neurilemma*.

The fibres which do not unite with each other in the
trunks lie parallel side by side. When, however, the
nerve trunks enter special organs, or when they approach
their terminations in the *skin*, the *muscles*, or elsewhere,
the nerve-fibres divide, becoming finer and finer until
they ultimately inosculate, or *terminate* in *loops* or other-
wise in a manner not yet worked out by *histologists*.
The finer *terminal* nervous filaments are not divisible
into *tube-sheath, axis-cylinder* and *contents*, but in the
present state of our knowledge seem to consist of *homogeneous* nerve substance.

Many of the terminal nerve-fibres which go to the
skin appear to terminate in *spiral coils* which embrace
or wind round the *tactile corpuscles*. (See *Tactile
Corpuscles*). Other nerve-fibres terminate in the middle

of the *Pacinian* corpuscles. The finer nerve-fibres of the brain and spinal cord often terminate in the caudate or stellate processes of the *nerve-corpuscles*.

148. Gelatinous **Nerve Fibre**. Besides the so-called *tubular nerve-fibres*, a second kind of fibre, consisting of smaller, softer, flattish, homogeneous yellowish-gray nerve filaments which contain *nuclei*, but do not *differentiate* in structure into sheath, axis-cylinder, &c., like those just described, are known to physiologists as *gelatinous nerve-fibres*.

The *olfactory nerves* and the nerves of the *sympathetic* or *ganglionic system* proper, consist mainly if not exclusively of *gelatinous nerve-fibre*. When these *fibres* are treated with acetic acid their *nuclei* become plainly visible.

149. **The Chief Property of Nerve-Fibre**, or as it has been termed its *neurility* or *excitability*, is its power of conveying or conducting *nervous force* or *stimulus* from the *cerebro-spinal axis* or a *ganglionic centre* by which a *muscle* is made to *contract*, or a *gland* to *secrete;* or by which external impression or irritation is transmitted from the skin, the organs of the senses, or any of the internal organs to the *brain* or *spinal* cord. Their function in this respect has been aptly compared with that of the *telegraph wires* in a *telegraphic system*.

The nerves, then, simply act as *conductors* of nervous *force* or *impression*. Helmholtz, Baxt, and Hirsch have estimated by means of induction (electric) currents made to act on a galvanometer, the *velocity* at which the nervous force or impulse *travels* in the motor nerves of a frog, at about 34 metres or 110 feet per second. It travels at different rates in different nerves, and the nerves of different animals.

150. **Ganglionic Corpuscles, Nerve Cells**, or **Vesicular Neurine** consists of minute spheroidal or stellate *nucleated cells* charged with finely *granular* matter containing a greater or less number of reddish or brown *pigment granules*, which give the *exterior* of the brain and *interior* of the spinal cord the peculiar pinkish-gray or *cineritious* appearance they present.

Nerve cells or *corpuscles* are of two kinds (1.) *Simple,* that is, having an *uninterrupted* outline, and being therefore spherical or ovoid, and more or less resembling in appearance glandular epithelial cells. (2.) *Caudate* or *stellate* (as shown in the diagram), that is, containing a number of tails or processes by which they become more or less *star-shaped.* (See fig. 35.)

The latter are termed *uni*-polar, *bi*-polar, *multi*-polar, according to the number of the processes they give off. These processes, which are *tubular*, contain granular matter. Some of them taper off to a point, others unite to the processes given off by other nerve-corpuscles, others again unite with the ends of ordinary nerve-fibre.

151. Function of the Nerve Corpuscles. The nerve corpuscles are very generally described as the *generators, sources,* or *originators* of *nervous force.* This statement however is only true with considerable modification.

They constitute the mass of the *exterior* of the *brain,* the *interior* of the *spinal cord* and of the substance of the *ganglia,* or the *nerve centres* at which *nervous force* is either *generated,* or nervous impressions are *radiated, transferred, diffused,* or *reflected.* (See *Brain,* &c.)

CHAPTER VI.

THE BLOOD.

152. The Blood, or *nutritive* fluid, is by far the most abundant and important fluid in the body: it has therefore been designated the " river of life."

Within the body in its *normal* or healthy state, it is entirely contained within a set of special vessels, termed *blood-vessels,* consisting of *arteries, veins,* and *capillaries.* It never escapes from these vessels into any other part of the body, except as a result of disease or injury, as in the case of *bruises* or *scurvy:* it is then said to be *extravasated.*

153. Appearance and Properties of the blood. The blood as it escapes from the body by a cut or injury, or when quite freshly drawn, *appears* to the unaided eye to be a *homogeneous*, somewhat viscid, bright *crimson* or *scarlet* fluid. It has a *saline* taste, is slightly *alkaline*, and is a little *heavier* than water, 1,055 parts by weight of *blood* being equal in *bulk* to 1,000 parts by weight of *water*. It has a slight peculiar *odour* or *halitus*, which is greatly increased by the addition of dilute sulphuric acid, containing about one-half of water. It is said that the blood of different animals may thus be readily distinguished by the *odour* evolved; this, however, is probably true of a few animals only.

About *one-half* of the blood contained in the body—that is that portion which is contained in the *veins* and in the *pulmonary arteries*—is of a *dark* purplish colour; this is termed *venous* blood.

Only that portion of the blood which is contained in the *arteries*, (and in the *pulmonary veins*), and freshly drawn blood which has been *exposed to the air*, possesses the bright *scarlet* colour usually deemed so characteristic of the blood. This is termed *aërated* or *arterial* blood.

If *dark venous* blood be shaken up in a bottle of *oxygen* gas, it will be immediately converted into bright *scarlet* blood, having all the properties of *arterial* blood. The same changes take place, though much more slowly, when blood is agitated in *common air*.

154. Magnified Blood. When a *very thin* layer of freshly drawn blood is examined by means of a powerful magnifying glass, it appears to consist of minute particles of a pale, *yellowish*, gritty substance floating in a *colourless* fluid. These minute gritty-looking bodies are the *corpuscles* of the blood.

To observe the *microscopic structure* of the blood accurately—

1. Take a slip of thin plate glass three inches long and one inch wide.
2. Twist a piece of thick string several times tightly round the middle of the last joint or end of the middle finger of the left hand.

3. Prick the end of the finger (which has now become purple and swollen because of the distension of the veins produced by the obstruction to the circulation caused by the ligature) with a sharp *clean* needle. A biggish drop of blood immediatley exudes, the operation causing little or no pain.

4. Smear a very *small quantity* of the blood thus drawn, so as to form a *very thin yellowish-looking layer* of blood on the middle of the glass slip. *Cover* immediately with a second plate of very thin glass.

5. Place the blood thus prepared under a microscope magnifying 300 diameters.

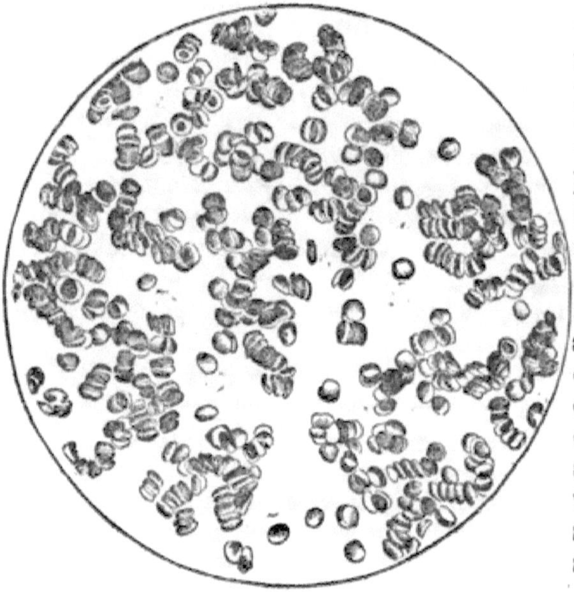

Fig. 36. Drop of Blood (Magnified).

On looking through the microscope, a picture, very correctly represented by Fig. 36, which is drawn from an actual photograph of the appearance of a drop of blood, only the $\frac{1}{140}$ of an inch in diameter, as shown under the microscope, will be seen.

155. Structural Character of the Blood.—The blood is thus seen to consist of (1) *red* and (2) *white* or colourless corpuscles; also, if examined under a much higher microscopic power, (3) of minute *oil globules*, floating in a nearly colourless transparent fluid, (4) the *liquor sanguinis* or blood plasma.

156. The General Composition of the Blood varies very considerably with the age, sex, temperament and health of the individual; also, with the time since the last

meal was taken, with the part of the body from which the blood is drawn, and with other conditions. Copious *bleeding* very considerably *reduces* the proportionate number of *red corpuscles.* Animal food, the use of certain medicines, as those containing *iron,* &c., very materially *increases* their number. In general there are proportionally more *red* corpuscles in the blood of *men* than of *women,* and of *strong* men than of *weak* men, and in the blood of young and middle-aged adults than in that of children and *old* people.

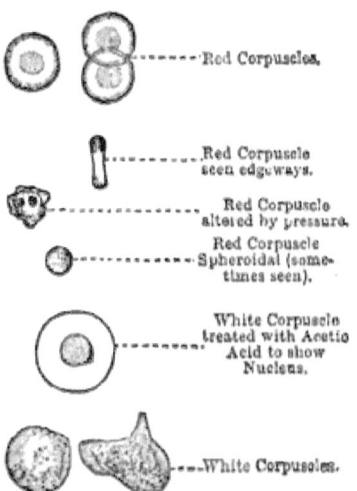

Fig 37. Human Blood Corpuscles.
Magnified 600 diameters.

One thousand parts of blood consist of about 500 parts of *liquor sanguinis,* and 500 parts of moist *red* and *white corpuscles.*

The average composition of human blood may be stated approximatively as consisting in each 1,000 parts of—*water,* 780 ; globulin, 140 ; albumen, 70 ; fibrin, 2 ; fatty matters, $1\frac{1}{2}$; extractive, $6\frac{1}{2}$.

COMPOSITION OF HUMAN BLOOD.

Water,					779·0
Solids,	Corpuscles,	Haematin,	7	141·0	
		Globulin,	134		
	Fibrin,			2·2	
	Albumen,			69·4	221·0
	Fatty matters,	Serolin,	0·2	1·6	
		Phosphorized fat,	0·49		
		Cholesterin,	0·09		
		Saponified fat,	1·00		
	Extractive matters,			6·8	1,000·0

The *liquor sanguinis* contains most of the *chlorine* and the *soda* of the blood, thus differing chemically from the corpuscles, which contain most of the *fatty* substances, and the *phosphates; all* the *iron*, and most of the *potash*.

157. The Red Corpuscles of human blood consist of exceedingly *minute*, soft, flexible, elastic, pale yellowish, circular, *biconcave, non-nucleated* discs, with rounded edges.

Their diameters vary from about the $\frac{1}{4000}$ to the $\frac{1}{3000}$ of an inch. Their thickness is about $\frac{1}{3}$ to $\frac{1}{4}$ of their diameter, and therefore varies from about the $\frac{1}{12000}$ to the $\frac{1}{9000}$ of an inch.

It has been roughly calculated that 10,000,000 red corpuscles would lay on one square inch of surface, and that 120,000,000,000 might be contained within the volume of one cubic inch.

It has also been estimated that one cubic inch of

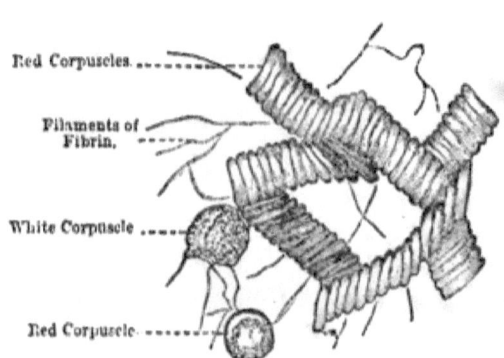

freshly drawn healthy human blood actually contains 84,000,000 of *red* corpuscles, and 240,000 *white* or *colourless* corpuscles. Dr. Draper, the celebrated American physiologist, has estimated that 20,000,000 of these red cor-

Fig. 38. Red Corpuscles of Human Blood arranged in Rouleaux.
Magnified about 600 diameters.

puscles are *born*, and 20,000,000 of them *die* per second, or with each beat of the heart.

158. Under the microscope these little bodies may be seen rolling and turning about in the *liquor sanguinis*, and arranging themselves in little piles, or *rouleaux*—like piles of small coin seen edgewise. When they *absorb oxygen* they become flattened, their walls becoming thicker and more opaque, and possibly more reflective.

When they absorb carbonic acid gas the cells are said
to become rounder and larger, and their walls thinner,
more transparent, and darker.

159. The Form and Size of the Red Corpuscles varies
in different animals.   They are *circular* and *biconcave*
in nearly all the mammalia, being smallest in the *deer*
tribe,—are *oval* in birds, reptiles, and fishes, being
largest in the reptiles.   The following diagram (fig. 39)

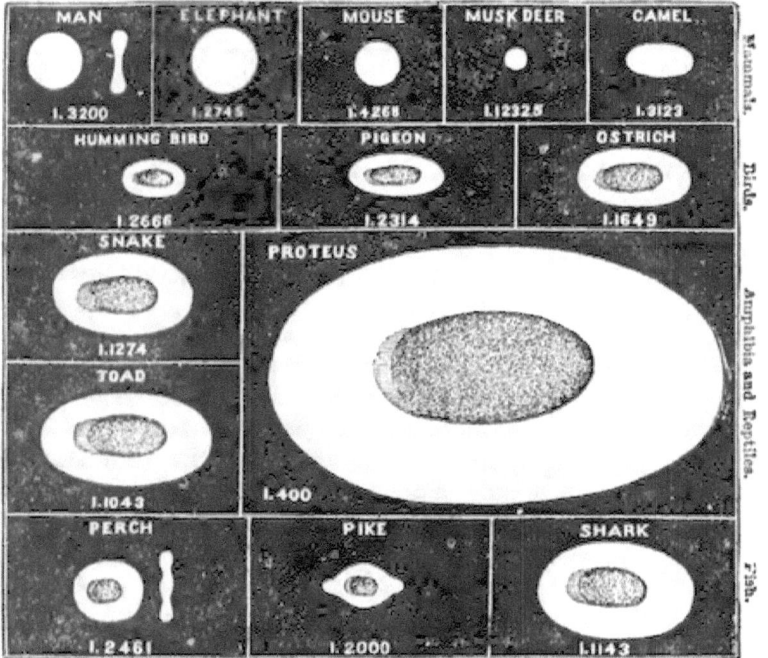

Fig 39. Shapes of Blood Corpuscles (red) of Various Animals.

gives their shape, also their longest diameters in fractions
of an inch.

160. The Structure of a Red Corpuscle is not even yet
positively determined.   It is usually described as consisting of an *interior semi*-fluid, or even quite fluid substance,
surrounded by an *outer* substance of gradually increasing
density, forming a not very definite or distinctly marked
cell wall.   Some physiologists still describe the red
corpuscle as a *homogeneous structureless* mass,

161. The interior substance of the corpuscle consists of *hæmoglobin*, or as it is sometimes variously termed *cruor*, or *cruoro-globulin*, or *hæmato-globulin*. This compound may be resolved into two substances:—*hæmatine*, the peculiar colouring substance of the blood, and an albuminous substance termed *globulin*.

Prepare a *second* glass slide (as described in section 154), but previous to covering the blood up with the thin covering glass, dilute with water. When examined under the microscope the red corpuscles will be observed to swell out their *sides*, losing their hollow, sunken, *dish-like* forms, becoming rounded and globular, the corpuscles ultimately bursting by *endosmose* or absorption of liquid.

When the *red corpuscles* are immersed in solutions of greater density than their own substance, they lose fluid by *exosmose*, shrivelling up, and losing their *normal* shape.

The red corpuscles have been variously termed *blood-cells, blood-corpuscles, blood-discs, blood-globules,* &c.

162. **The Colour of the Blood** is due to the *cruorin* or *hæmatin*, as it is termed, which is contained in the red corpuscles. This substance exists in two states, viz.:—that of *purple* cruorin, in which it is combined with *less* oxygen, and *scarlet* cruorin, in which it is combined with *more* oxygen.

Venous blood being *deficient* in oxygen is now supposed to owe its *dark purple* colour to the presence of the *lower* oxide of cruorin.

Arterial blood being supplied with *excess* of oxygen, is supposed to owe its *bright scarlet* colour to the presence of the *higher* oxide of cruorin.

That the colour of the blood is not due, as was formerly supposed, to the presence of *iron*, is shown by the fact that *hæmatin* still retains its peculiar *red* colour after all its *iron* has been extracted.

That the *dark* colour of *venous* blood is simply an *optical* effect due to the greater *thinness* and consequent *transparency* of the cell wall of the *venous* corpuscle, is also a theory which is now being generally abandoned by physiologists.

163. **The Detection of Blood Stains** in case of murder may, in general, be effected with great certainty by means of the *microscope* and the *spectro-microscope.*

Human blood may readily be distinguished not only from common dark or red paint, but also from that of many other animals by a clever microscopist, by means of the *appearance, shape,* and *size* of the discs, or *red corpuscles.* (See fig. 39.)

The *spectro-microscope* consists of a peculiar combination of glass *prisms* attached to the microscope, by means of which an object may be examined with the aid of special light obtained from any specific part of the *spectrum.*

A minute speck of blood, weighing not more than the *one-millionth* part of a grain, will, when examined under the *spectro-microscope,* give certain well marked characteristic dark lines, called *absorption bands,* due to the presence of *cruorin.* By suitable treatment the *cruorin,* notwithstanding the exceedingly minute portion present, may be changed respectively from the *higher* to the *lower* oxide, or *vice versa,* the well-marked character of the *absorption bands* varying accordingly.

164. **Function of the Red Corpuscles.**—The chief *function* of the *red corpuscles* of the blood is to act as "*carriers* of oxygen." They *absorb* large quantities of *oxygen* in the *lungs,* and *carry* and *deliver* it over to the *tissues,* even in the most remote parts of the body, thus *vivifying* them and enabling them to perform their various functions.

Blood *deprived* of its red corpuscles will only absorb 1-13th of the quantity of oxygen it dissolved in its *normal* state. Blood in this state is quite useless for all purposes of *transfusion* or vivification.

165. **The Oxygen in feeble Chemical Union with the Blood.**—When *pyrogallic acid* is exposed to the oxygen of the air, it readily enters into *combination* with it; but when it is mixed with the blood, it does *not* combine

14 E   F

with its absorbed oxygen. It has, therefore, been supposed by some physiologists that the oxygen in the
blood is not held in a mere state of *mechanical solution*,
but is held in a state of loose *chemical combination* by
some substance in the corpuscle.

166. The White or Colourless Corpuscles of the blood
are minute pearly, grayish, semi-transparent, spheroidal,
contractile, roughish-looking, nucleated bodies, about
the 1-2,500th of an inch in diameter (See fig. 37). In a
state of health the blood contains 400 to 500 times as
many *red* as *white* corpuscles.

The *white corpuscle* consists of an outer *cell wall*
(which bursts when placed in water), of a fluid containing more or less granular matter, and a *nucleus*.
The *nucleus* is rendered more distinct by immersing the
corpuscle in dilute *acetic acid*. (See fig. 37.)

167. When the white corpuscles are carefully watched
for a lengthened period as they float in the blood-*plasma*,
they are observed to undergo considerable change of
form, stretching portions of themselves out in the form
of *processes* or arms, in first one direction, and then in
another, thus manifesting a sort of independent life of
their own, and very closely imitating the nature, properties, and behaviour of the *amœba*, one of the very
lowest forms of animal life.

168. Functions of the White Corpuscles.—The functions of the white corpuscles are only partially determined, but they are supposed—

1. To develop the *red* corpuscles from their *nuclei*.

2. They are supposed to assist in developing the *fibrin*
of the *liquor sanguinis*.

169. The Liquor Sanguinis, or Blood Plasma, is the
clear, transparent, colourless, or slightly yellow and
slightly viscid, saline, *albuminous* liquid in which the
red and *white corpuscles* float.

It consists chiefly of a very dilute solution of albumen,
fibrin, fatty matters (serolin), and certain salts, chiefly
sodium chloride (common salt), and tribasic phosphate
of soda.

1,000 parts of *liquor sanguinis* contain 60 to 70 of albumen, and 2 to 3 of *fibrin*.

The phosphate of soda probably aids it in *dissolving* and holding the *carbonic acid* contained in the blood.

170. Functions of the Liquor Sanguinis :—

1. The *liquor sanguinis* floats the *red* and *white* cor·puscles of the blood, thus conveying them to all parts of the body.

2. The *liquor sanguinis* permeates the walls of the capillaries; and after thus escaping, irrigates and bathes the tissues, which appropriate the *nutritive material* they require. This (nutrition of the tissues) is probably its *chief* function.

3. It receives the products of the waste tissues and distributes them to the various *organs of excretion*, by which they are removed from the system.

171. Coagulation of the Blood (from Lat. *coagulo*, I curdle).—Let the student go to the butcher, and get him to collect some blood in a small basin in his presence, and let him watch the changes it shortly undergoes.

1. In from three to ten minutes the whole mass of blood in the basin *sets* into a soft, red, gelatinous, *homogeneous-looking* mass.

2. Shortly afterwards, small drops of a transparent yellowish fluid (the *serum*) begin to ooze out on its upper surface.

3. The *gelatinous red* mass, now gradually contracting, *squeezes* out the *serum* on all sides, until the bulk of this fluid becomes two to three times the bulk of the solid red mass which now, having much greater firmness, constitutes the *clot, cruor*, or *crassamentum* of the blood.

172. The *coagulation* of the blood thus comprises essentially *two* distinct processes : *first* a *setting* of the blood ; *second*, a sort of rough analysis, or separation of the blood into *serum* and *clot*.

173. The following Table shows the comparative composition of *living* and *coagulated* blood :—

Living blood, . .	Liquor sanguinis,	Serum. Fibrin.
	Corpuscles, . .	Red. White.
Coagulated blood,	Serum, ...	Water. Albumen. Salts.
	Clot, ...	Red Corpuscles. White ,, Fibrin.

174. Serum (from Lat. *serum*, whey).—The serum of the blood is the pale yellowish, slightly viscid, greasy *albuminous fluid* which is *squeezed* out from the clot during the process of coagulation. It consists of the liquid portion of the *blood-plasma*, from which the *fibrin*, or so much of the *fibrinogen* as is capable of giving rise to its *formation*, has separated by spontaneous coagulation.

175. The Serosity (from Lat. *serum*, whey) of the blood is the feeble aqueous solution of the *salts* of the *serum* which separates from the *serum* when it is coagulated by *heat;* in other words, the *serosity* of the blood is the incoagulable portion of the serum. The serosity also contains minute quantities of *extractive* and *fatty* matters.

176. Wounds inflicted a few hours *after* death may readily be detected by the absence of clot about the mouth of the wound.

Bleeding to death is often prevented by nature's stopping the mouths of the injured blood-vessels with *plugs* of *clot* or coagulated blood.

The coagulation of the blood also stops *internal* bleeding, and promotes the *healing* of incised wounds by joining their adjacent surfaces together.

When *pus* and some other substances get into the blood, they sometimes cause death by setting up the process of coagulation, by which *plugs of clot*, which stop the circulation, are formed in the principal arteries.

177. Conditions which Accelerate Coagulation.

Moderate warmth.

Rest.

Contact with solid bodies, especially those with rough surfaces.

Free exposure to air.

Shallow vessels.

The addition of a limited quantity of water.

Conditions which Retard Coagulation.

Contact with the walls of living blood-vessels.

A temperature of above 120 deg. or below 40 deg. F.

Deep vessels.

Exclusion from air.

The addition of alkaline solution, or of salts of the earth and alkalies; also by strong acids.

Death by suffocation from carbonic oxide, carburetted hydrogen, &c.

178. Functions of the Blood.—The following are the chief functions of the blood :—

(*a*) It *nourishes*, builds up, renovates or repairs, and vitalizes or revivifies the various tissues.

(*b*) It conveys oxygen to the various tissues, thus supplying the agent by whose chemical union with these tissues the heat, and *nervous*, *mental*, and other *vital* forces are developed in the body.

(*c*) It warms and moistens all parts of the body.

(*d*) It receives the *refuse*, the liquified product of the *oxidized* or waste tissues of the various parts of the body, and conveys them to the various *excretory* organs by which they are *eliminated* from the system.

(*e*) It supplies the various *glands* with the material out of which they *elaborate* or *secrete* the fluids or other substances necessary for the proper performance of the function of digestion, and of the other *functions* of the body.

179. Gains and Losses of the Blood.—During the course of its circulation the blood is *continually* losing and *gaining* material.

It *gains* matter, viz., *oxygen*, as it passes through the lungs; *waste products* (from the *tissues*) as it traverses the capillaries; exuded *blood-plasma*, and probably other substances, from the lymphatics; *sugar* (glucose), and possibly *white corpuscles*, from the liver; also, possibly, *white corpuscles* from the *spleen* and *ductless glands.*

The blood *constantly loses* matter:—*carbonic acid, aqueous vapour,* and *urea* by the *lungs;* water, *urea,* &c., by the *kidneys;* the materials of *bile* by the *liver;* the materials *selected* by the *tissues* for their repair and renovation; *aqueous* vapour, *carbonic acid,* and *urea* by the *skin.*

The blood constantly *gains heat* in *itself,* and from the *tissues* (by oxidation and combustion); but it constantly *loses heat* on the *free surfaces* of the body, chiefly by *radiation* and *evaporation.*

In addition to the above, the blood *gains* matter *intermittently* from certain sources, viz., oxidized or waste tissue; products from the muscles; liquified nutriment from the alimentary canal; water, &c., absorbed by the *skin.*

The blood also *loses intermittently* by many of the secreting glands, as the salivary glands, &c.

For difference between Blood and Lymph (see section on *Lymph*).

CHAPTER VII.

CIRCULATION OF THE BLOOD.—THE HEART AND BLOOD-VESSELS.

180. General Purpose of the Circulation.—The *Blood* constitutes the ever-flowing river of the human system. Its presence in a state of purity, at all points of the organism—to *render up* the *vitalizing* oxygen and the elements of *nutrition* and *repair;* also, to receive and remove the *poisonous* products of combustion, disintegration, and waste, so that they may be either utilized or expelled from the system, as the case may require—constitutes an ever-recurring necessity for the perpetual movement or circulation of the blood from the first dawn of life to its ultimate close. We now proceed to discuss and explain the nature of the *mechanism* and the *forces* which by this movement of *circulation* is initiated and sustained.

181. Circulation (from Lat. *circulo*, I encompass) is the process by which the blood is driven *out* from the heart, conveyed by the arteries and *arterial capillaries, to* the various parts of the body, and again returned *to* it by the *venous capillaries* and the *veins*, the whole forming one continuous movement or journey.

The "circulation" is sometimes described under the terms *greater* or *systemic* circulation, *lesser* or *pulmonary* circulation, and a *subordinate* branch of the systemic circulation, termed the *portal* circulation.

The *greater circulation* is that by which the blood is sent out from the *left* side of the heart, distributed by the *arteries* through the system, and returned by the *veins* to the *right* side of the heart.

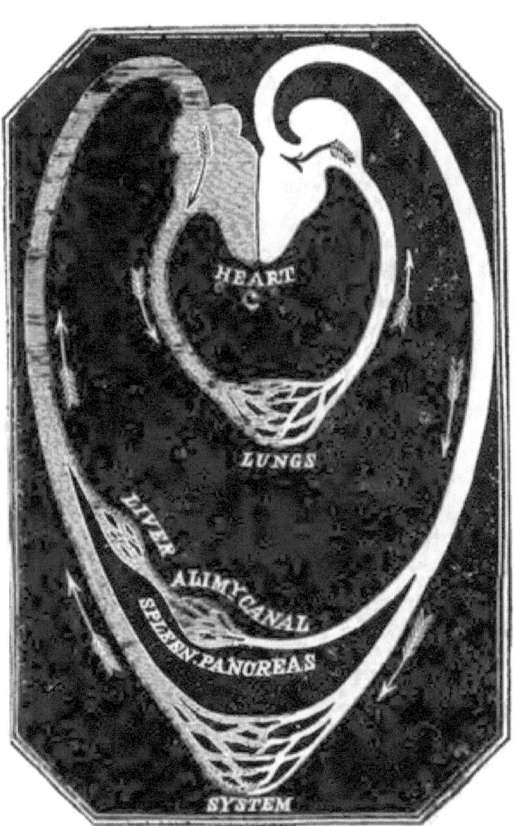

Fig. 40. Theoretic Circulation.

The *lesser circulation* is that by which the blood is

sent out from the *right* side of the heart to the lungs,
and returned from them to its *left* side.

The *portal circulation* comprises that portion of the
systemic circulation by which the blood, distributed to
the stomach, liver, spleen, pancreas, and the intestines,
is collected by the *vena porta*, and distributed through
the substance of the liver, on its way to the inferior
vena cava, which it reaches by the *hepatic* veins.

The *heart*, the *arteries*, the *veins*, and the *capillaries*
constitute the chief *organs of circulation*, and are designated the *"circulatory system."*

182. Position and Size of the Heart.—The heart is
situated in the *mediastinum* or central part of the
cavity of the thorax, immediately under the lower
two-thirds of the sternum or breast-bone. It is
almost entirely surrounded by the two lungs. Its
base is directed backwards and upwards, being supported chiefly by the great blood-vessels. Its apex,
or lower end, is turned *forwards* and *downwards*,
pointing to, obtruding upon, and to a certain extent
displacing the left lung; its *apex*, being placed nearly
opposite the interval between the fifth and sixth ribs, may
be felt by the hand striking against the walls of the
chest during the beating of the heart. (See figs. 2, 3,
and 41 and 43.)

The heart is about 5 inches long, $3\frac{1}{2}$ inches broad, and
$2\frac{1}{2}$ inches thick. Its average weight in the male is ten
to twelve ounces, in the female eight to ten ounces.
It forms about $\frac{1}{160}$ of the weight of the whole body of
the male, and $\frac{1}{140}$ of that of the female.

183. The Heart, which is a kind of *force-pump*, is the
principal organ of circulation. From it the blood
acquires the *propulsive* force by which it performs the
various movements just described.

The heart is a hollow, conical, fleshy bag, about the
size of a man's fist. It consists of involuntary but
striated muscular fibre. The *heart* is divided by *septa*
and *valves* into *four cavities*, which have no *direct* com-
munication with each other; that is, the blood on one

side of the heart cannot pass over to the *cavities* on the other side of the heart, without passing through the blood-vessels in the lungs.

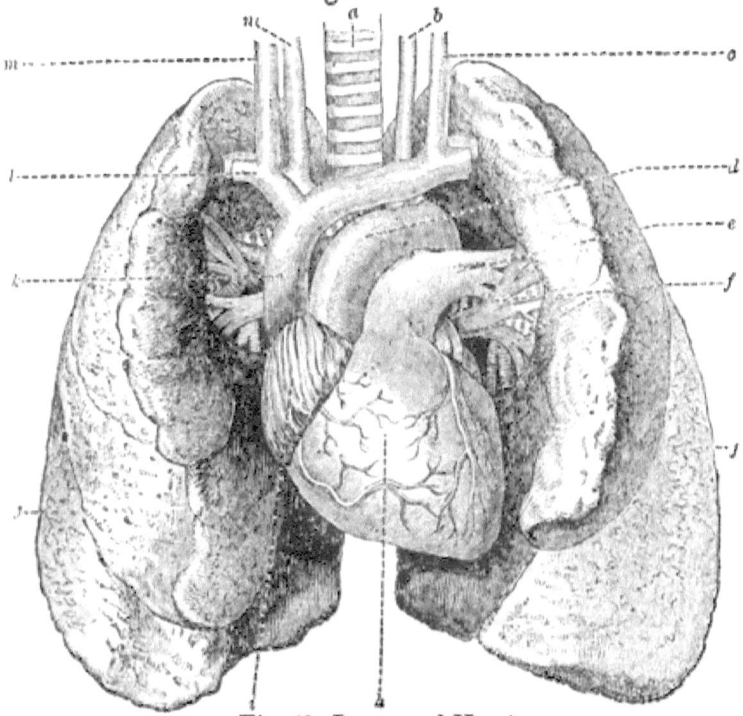

Fig. 41. Lungs and Heart.

a, Top of Trachea. *b,* Left Carotid Artery. *c,* Left Jugular Vein. *d,* Arch of Aorta. *e,* Pulmonary Artery. *f,* Bronchi and Blood-vessels. *g,* Left Lung. *h,* Right Ventricle. *i,* Right Auricle. *j,* Third Lobe of Right Lung. *k,* Superior Vena Cava. *l,* Right Subclavian Vein. *m,* Right Jugular Vein. *n,* Right Carotid Artery.

The heart is completely enveloped in a closed sac of *serous* membrane, termed the *pericardium.*

The human heart is usually described as containing two distinct sides (a right and a left), separated by a fleshy wall, termed the *median septum* of the heart. Really the heart of man, of birds, and of the mammalia may be said to consist of two complete hearts—a right and a left heart—each heart corresponding to the single complete heart of a fish; the human heart thus forming a *double* heart, and the circulatory movement set up by it a double circulation, termed respectively the *greater*

or *systemic* circulation, and the *lesser* or *pulmonary
circulation.*

The right and left sides of the heart each contain two

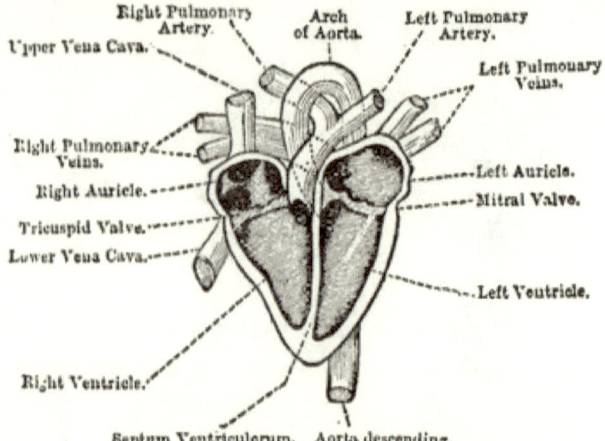

Fig. 42. Theoretical Section of the Human Heart, seen
from the front.

cavities—viz., an *upper* one with *thinner* walls, termed
an *auricle;* and a *lower* one with *thicker* walls, termed
a *ventricle.*

184. **The Auricles,** or *upper* chambers of the heart, are
separated from, and open into, the *ventricles* by means
of transverse constrictions, each of which is strengthened
by a fibrous ring, termed the *zona annularis:* these form
the auriculo-ventricular openings, to the edges of which
the valves of the heart, the *mitral* and the *tricuspid*
valves, which close or open these cavities, are attached.

185. The *walls* of the auricles are comparatively *thin;*
the passage of the blood into the ventricles depending
less upon the *propulsive* power exerted by the walls of
the auricles, than upon the *suction* consequent upon the
dilatation of the cavities of the ventricles.

The right auricle communicates directly with the *superior* and *inferior* venæ cavæ ; the *inferior vena cava*
only being protected by a valve—the *Eustachian valve.*
The orifice by which it communicates with the *right ventricle* is guarded by the *tricuspid* valve. (See fig. 43.)

The *left auricle* communicates directly with the four

pulmonary veins which return *purified* blood from the

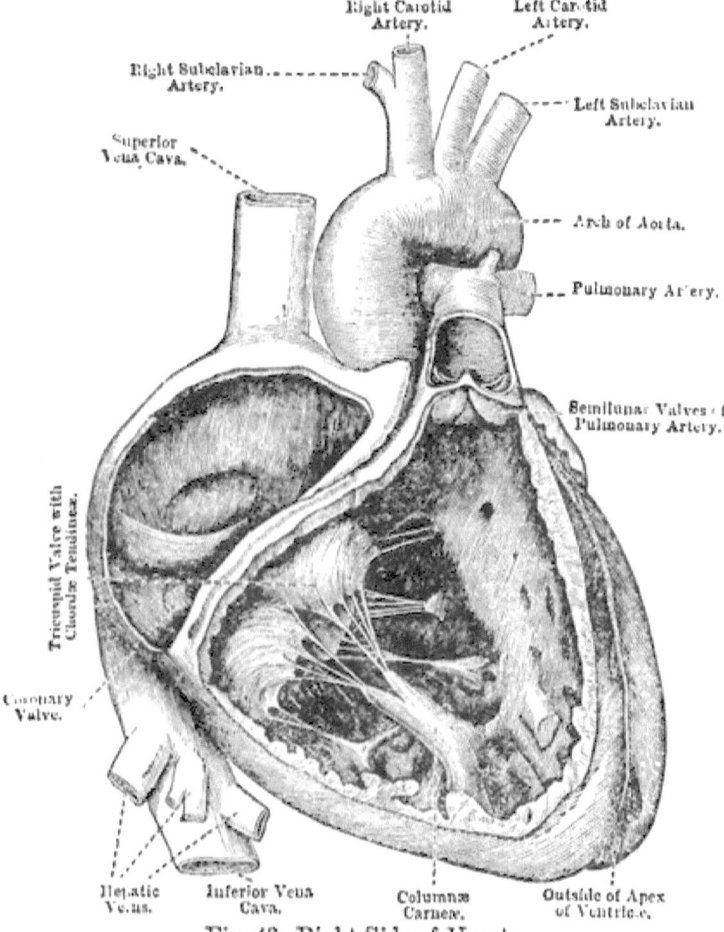

Fig. 43. Right Side of Heart.

lungs. Its lower opening, by which it communicates with the left ventricle, is guarded by the *mitral valve*.

186. The Ventricles (from Lat. *venter*, stomach) are the two lower and *thicker* walled cavities of the heart. They are separated from each other by the ventricular portion of the *median septum* of the heart. The walls of the *left* ventricle are much thicker than those of the *right*. Each ventricle is capable of containing four to five ounces of blood. The ventricles are a little larger than the auricles.

The *pulmonary artery*, the lower end of which is protected by semi-lunar valves, opens into the *right ventricle*. When the right ventricle *contracts*, the tricuspid valve *closes*, and the *venous blood* is driven out of the ventricle through the pulmonary artery into the lungs, to be oxidized.

187. The *left ventricle* opens into the *aorta*, the entrance of which is protected by the *aortic* (semi-lunar) *valves*. When this *ventricle* contracts, the *mitral valve* closes, and the blood is *propelled* through the *aorta*, and by it is transmitted to the general system.

As the *muscular force* required to drive the blood through the general system is much greater than that required to propel it through the *lungs* only, the walls of the left ventricle are much more largely supplied with *muscular fibre*, or in other words, are *thicker* than those of the *right* ventricle.

The *columnæ carneæ* are rounded muscular columns which project from the sides of the ventricles (see fig. 43). They are well shown on the inner surface of the ventricles of a sheep's heart.

The *columnæ carneæ*, to which the minute tendinous cords (*chordæ tendineæ*) are anchored, are termed *columnæ papillares*. (See fig. 43.)

188. The *chordæ tendineæ* are the minute white tendinous cords (shown in the diagram), by which the valves of the heart are anchored so as to close up the auriculo-ventricular openings. Were these cords to be cut or broken, the flaps of the valves would be carried through into the auricles on the contraction of the ventricles, and thus become useless.

As the distance from the walls of the ventricles to the valves varies during their contraction, the length of the *chordæ tendineæ* remaining uniform, some adjustment is needed to adapt them to the required length; this adjustment is effected by the *carneæ columnæ* (fleshy columns), to which their lower extremities are attached.

189. Movements of the Heart.—When the two auricles become charged with blood, the right with *venous* blood from the *venæ cavæ*, the left with *arterial* blood from the *pulmonary veins*, they gradually and simultaneously contract, driving their contents into the ventricles. This contraction constitutes the *systole* of the auricles.

When the ventricles, which dilate to receive the blood from the auricles (the valves being open), have become thus charged with blood, after a short pause they begin to *contract*, and thus simultaneously both close their valves and propel their contents—those of the right auricle into the *pulmonary artery*, and thence to the *lungs*—those of the *left ventricle* into the aorta, and thence to the general system, as previously described.

These contractions constitute the *systoles* (from Gr. *sustello*, I contract) of the ventricles.   The contractions or *systoles* of the auricles and ventricles take place alternately—that is, while the *ventricles* are *contracting*, the *auricles* are *dilating*.

The dilatation or expansion of these cavities is termed their *diastole* (from Gr. *diastello*, I expand).

190. The *beating of the heart*, felt by the hand placed over the chest, is produced by the *systole* of the ventricles, which suddenly causes, from the peculiar arrangement of the muscular fibre of the heart, the apex to bend upwards so as to kick against the side of the chest, this movement being increased by the stretching and elongation of the *aorta* by the blood suddenly forced into it.

The movements of the heart thus take place in the following order:—

(*a.*) The walls of both *auricles* contract simultaneously.

(*b.*) Immediately afterwards the two *ventricles* contract simultaneously.

(*c.*) Then comes a *pause* of much longer interval than that occupied by these double contractions, after which the movements are repeated as before.

The auricles are *dilating* all the time the ventricles are *contracting*.

191. Course of the Blood through the Heart. (See next page, fig. 44.)

1. The *venous* blood enters the *right* auricle through the *superior vena cava*, and the Eustachian valve of the *inferior vena cava*.

2. It then passes (the Eustachian valve having closed) through the auriculo-ventricular opening (the passage between the *tricuspid valve*) into the *right ventricle*.

3. It then passes (driven by the contraction of the right ventricle and the closure of the *tricuspid valve*) through the *pulmonary arteries* into the *lungs*.

4. It is poured *back* (oxidized and purified) by the four *pulmonary veins* into the *left auricle*.

5. The *left auricle* now contracts, and drives the blood through the *left auriculo-ventricular* opening (the *mitral valve* being open) into the *left ventricle*.

6. The *left ventricle* now contracts, the *mitral valve* simultaneously closing, while the aortic valve opens, and drives the blood through the *aorta* into the general system.

192. Experimental Proof of the Course of the Blood through the Heart.—Procure a sheep's heart, having instructed the butcher *not* to cut off the blood-vessels within a few inches of the heart itself. Tie up one of the *venæ cavæ*—insert a glass tube into the other—inject

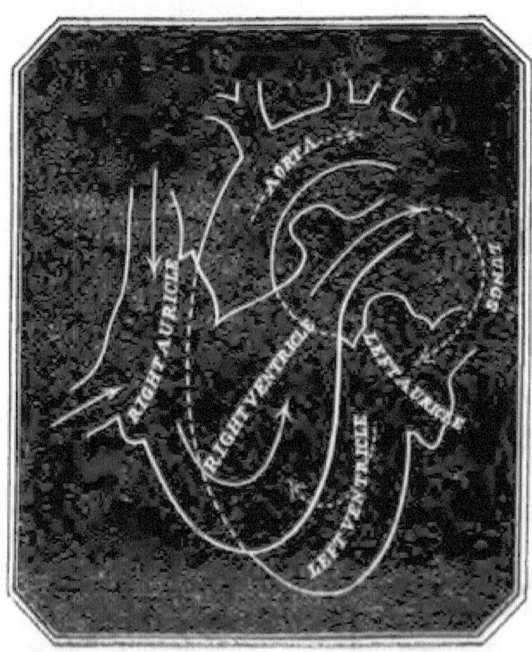

water into the heart through the tube, it will flow out of the *pulmonary arteries* and will not reach the *left* side of the heart.

Repeat the experiment by injecting water through one of the *pulmonary veins* (the other being tied), the water will flow out of the *aorta*, and not through the *right* side of the heart.

Fig. 44. Diagram of the Circulation through the Heart (after Dalton.)

If the *tops* of the ventricles be cut off and the ventricles filled with water, the aorta and pulmonary veins being tied, the thin membranous valves will be pressed upwards, becoming tightly stretched, and the whole action of the valves clearly visible.

193. Valves of the Heart (from Lat. *valvæ*, folding doors).—The valves are tough, flexible, *membranous* structures, which are attached to the fibrous borders of the openings into the heart, and *between* its upper and lower chambers. (See figs. 43 to 45.)

They are so arranged as only to permit of the passage of fluid in one direction, any attempt at *reflux* causing the blood to get *behind* the flaps or segments of the valves, and thus force them tightly against the openings, so as to close them the more effectually in proportion as the backward pressure is increased. But for the *chordæ*

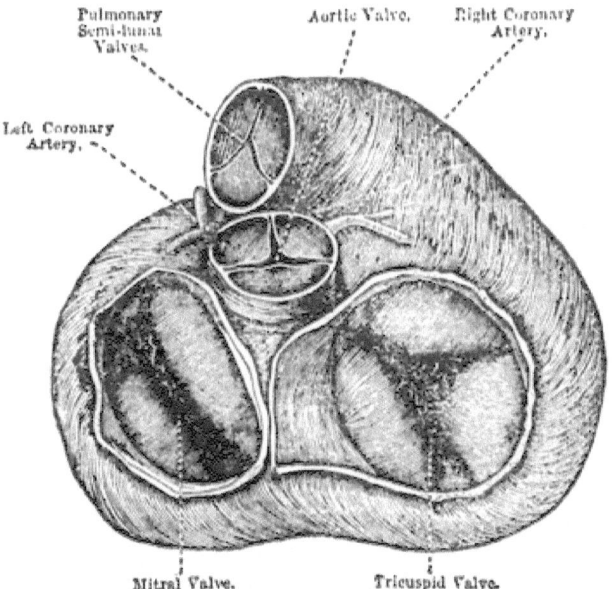

Fig. 45. The Top of Heart, the Auricles being dissected off.

tendinæ, the pointed flaps of the tricuspid and central valves would be driven through *into* the *auricles*.

The valves which consist chiefly of connective tissue,

are formed by the duplicature of the *lining membrane* of the interior of the heart, their substance being strengthened by an additional layer of *fibrous tissue*, possibly containing muscular fibre.

194. The heart is supplied with *six* valves, *four* of the principal of which, viz., the *tricuspid* and the *pulmonary* semi-lunar valves of the right side of the heart, and the *mitral* and *aortic* (semi-lunar) valves on the left side, are shown in fig. 45. The right auricle also possesses two other valves, viz., the *coronary valve*, which guards the entrance of the *coronary* vein (see fig. 43), and the *Eustachian valve*, which protects the termination of the inferior vena cava.

195. The *tricuspid valve* (from Lat. *tres*, three ; and *cuspis*, a point), so called because it consists of three pointed membranes, when closed, separates the right auricle from the right ventricle. Its structure, general arrangement, and connection by the chordæ tendineæ, and its mode of action, are clearly shown in fig. 43.

196. The mitral, or *bicuspid valve,* so called from its fancied resemblance to a bishop's mitre, consists of *two pointed* membranes. Its general structure and mode of action resemble those of the tricuspid valve. When closed it separates the *left auricle* from the *left ventricle.*

197. The semi-lunar valves of the *aorta* and of the *pulmonary* artery, and the *Eustachian* valve of the inferior vena cava, consist of three half-moon or crescentic shaped membranous folds or pockets, the *convex* borders of which are *attached* to the sides of the blood-vessels, their straighter edges turned towards the centres of the vessels being *free*, and pointing in the direction. in which the current flows.

When the current is passing in its normal direction, the blood flows between the free edges of these membranous pouches, pushing them close up against the walls of the vessels, and thus making a clear passage for itself. When, on the contrary, the current attempts to *return*, the blood forces its way into the pouches behind the valves, and fills them out, pushing their free edges together, thus raising a partition or obstruction in the middle of the tube, which prevents all further movement of the blood in that direction. (See figs. 42 to 45.)

In order the more effectually to close up the *central* spaces between the valves, the middles of the *free edges* of these valves are frequently furnished with little nodular bodies, termed the *corpora Arantii.*

198. The Arterial Pulse (from Lat. *pulso,* I beat) is the alternate *swelling* and *contracting*, and consequent *beating* of the arteries, which is felt when the finger is placed on the arteries of the temple, wrist, ankle, or other part of the body.

The pulse is caused by the *systole* or contraction of the *ventricles* of the heart, which drives an additional quantity of blood into vessels already *quite full;* they consequently become distended and enlarged both in *diameter* and in *length.*

199. The Vigour of the Pulse becomes *feebler* the more *remote* it is from the heart until the blood has passed through the capillaries into the veins, in which it entirely disappears. This gradual diminution of the force of the pulse depends—

(*a.*) On the *resistance* caused by *friction.*

(*b.*) On the *greater space* contained in the *small* as compared with the *larger* blood-vessels.

The arterial system has been compared, from this point of view, to a *cone,* the *base* of which, consisting of the smaller arteries and capillaries, is turned away from the heart, that is, towards the periphery of the body. But it has also been compared, perhaps more correctly, to a *tree,* the smaller or more remote branches of which occupy a much greater cubic space than that occupied by the trunk itself.

The cessation of the pulse consequent on the blood's passing through a large number of small tubes, the combined *sectional area* of the interior of which exceeds that of the original vessel by which they are supplied, may

14 E   G

be easily demonstrated practically by the following simple experiment:—Get a small *force pump*, or good sized injection syringe, a yard or so of elastic (india rubber) tubing, a piece of sponge, and a funnel. Fix the sponge securely, so as to close up the large end of the funnel; insert the small end of the funnel in one end, and the injecting syringe into the other end of the vulcanized tubing. Fill the tube with water, and then continue to inject fresh quantities of the liquid by means of the pump or syringe. The student will observe that every additional stroke of the pump, after the tube is filled with fluid, will give rise to a swelling and "kicking" of the tube (the latter due to its elongation), thus imitating the passage of the blood through the artery, with its accompanying phenomenon, the pulse. The water, however, will continue to flow evenly and uniformly, that is, without any *jerking* after passing through the pores of the sponge, thus also imitating the passage of the blood through the *capillaries* into the veins.

200. The Nerves of the Heart are derived from three sources :—

(1.) It possesses certain *intrinsic ganglia*—that is, *ganglia* lodged in its own substance. These, in all probability, give it its power of *rhythmic movement.*

(2.) From the *sympathetic* system, the influence by which the heart is excited to increased action by joy or other emotion, is in all probability conveyed to it by these nerves.

(3.) It receives branches of the *pneumogastric*, or *tenth* pair of nerves, which come directly from the brain. These latter exert an *inhibitory* influence on it, probably stopping its action, as in certain cases of death from *fright.*

201. The Arteries (from Gr. *aer*, air; and *tereo*, I keep) are the blood-vessels by which the blood is carried out from the heart, and distributed to the lungs and to the rest of the system. As the larger part of the blood contained in the arteries consists of pure, *oxidized*, scarlet blood, intended to nourish and vivify the general tissues of the system, it is generally termed *arterial blood*. The *pulmonary arteries*, on the contrary, contain *venous* (dark-coloured) blood, which they convey *to* and distribute through the vessels of the lungs, for the purpose of purification.

The *arteries* commence in one large vessel, the *aorta* (see figs. 40, 42, and 46), which divides and subdivides into

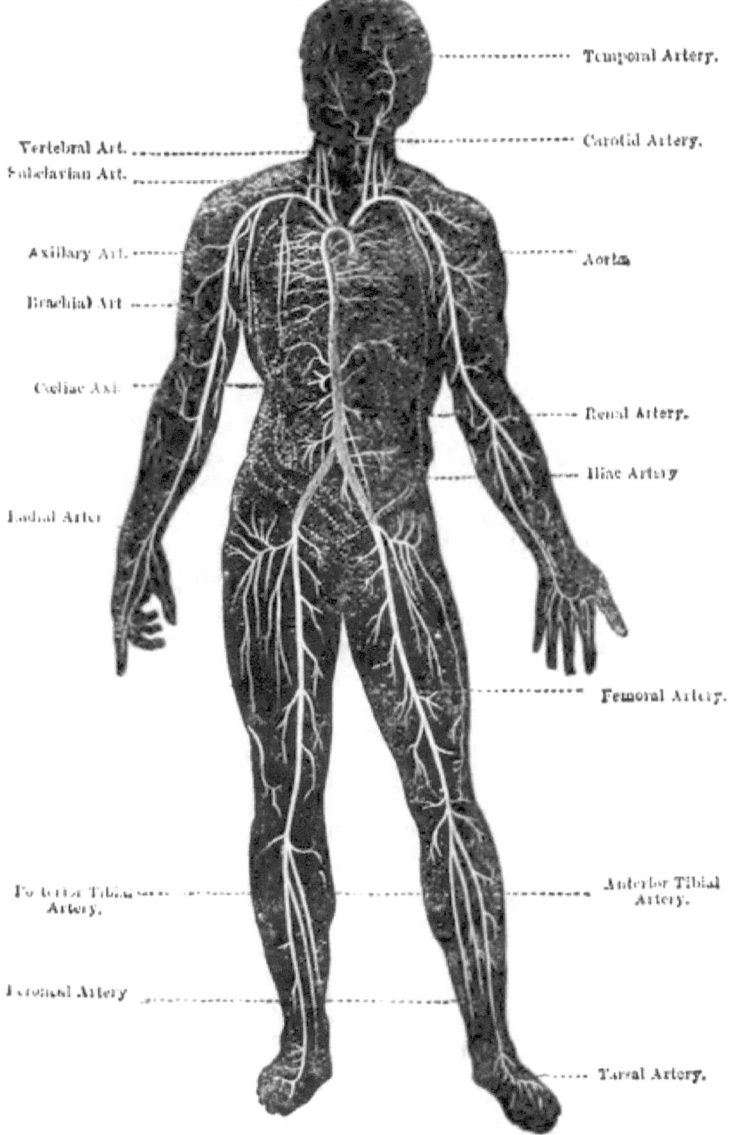

Temporal Artery.

Vertebral Art.
Subclavian Art.

Carotid Artery.

Axillary Art.

Aorta

Brachial Art

Coeliac Axis

Renal Artery.

Iliac Artery

Radial Arter

Femoral Artery.

Posterior Tibial Artery.

Anterior Tibial Artery.

Peroneal Artery

Tarsal Artery.

Fig. 46. Showing Arterial System of Man.

a large number of branches, which become, like those of a tree, more and more minute as they proceed farther from the trunk, until they ultimately *terminate* in the *capillaries*.

202. The *arteries* derived their name from the fact of their having been supposed by the ancients to contain air, being generally found empty after death.

203. The chief arteries are the *aorta*, or *systemic artery;* the *innominate* arteries, which send off the carotid arteries to the head, and the *subclavian*, which supply the *axillary*, the *brachial*, the *ulnar*, the *radial*, the *palmar*, and the *digital* arteries with blood; the *iliac* arteries (external and internal); the femoral arteries, which supply blood to the lower limbs; the mesenteric and *renal* arteries; and the *cœliac axis*, which gives off the *gastric*, *hepatic*, and *splenic* arteries. (See fig. 46.)

The leading arteries are distributed through the body in the general direction of the chief bones of the *skeleton*, from their proximity to which they frequently derive their names, as in the cases of the *temporal, femoral, occipital, radial*, and *ulnar* and other arteries.

Their safety is secured by the protected situations they occupy, being in general *deeply* placed *near* and on the *flexor* sides of the bones, passing up the centres of the limbs, or in central positions in the great cavities of the trunk.

204. **The Aorta** (from Gr. *œiro*, I suspend) is the *main trunk* of the arterial system (See figs. 42 and 43). It consists of a tough, flexible, cylindrical tube, rather less than one inch in diameter, which arises from the *upper* part of the *left ventricle*, and after proceeding *upwards* for about two inches, arches backward to the *left side;* and passing over the root of the left lung, descends through the *thorax*, on the left side of the vertebral column to the *diaphragm*, through the aortic opening of which it passes into the cavity of the *abdomen*, where, considerably reduced in size, it terminates opposite the *fourth lumbar* vertebra. At its termination it divides into two branches, which form the right and *left* common *iliac* arteries. The *aorta* is thus divisible into the *arch*, the *thoracic*, and the *abdominal* aorta.

205. The Cœliac Axis is a short *trunk* artery, about half an inch in length, which projects horizontally from the *aorta*, just opposite the *aortic opening* of the diaphragm. It divides into three branches, constituting respectively the *gastric, hepatic,* and *splenic* arteries (see fig. 46), which supply the stomach, liver, and spleen with arterial blood.

206. Structure of the Arteries.—The walls of all but the smallest arteries consist of three coats, viz.—(1.) an *internal* or *epithelial* coat; (2.) a *middle* or *contractile* coat; (3.) an *external* or *areolar* coat.

207. The *arteries* and the *veins* may readily be distinguished from each other in the dead body, as in the joints of meat which come from the butcher; the former (the arteries) being *round*, and having their *walls* comparatively stiff and thick; while the *walls* of the *veins* are *collapsed* and *flaccid*. In general, an artery cut, even when empty, preserves its cylindrical form; whereas the veins *collapse* under similar circumstances.

In the *smaller* arteries, just above the size of the capillaries, this *threefold structure* disappears; the walls of these arteries consisting of nearly homogeneous almost transparent membrane, containing *muscular fibre*.

208. The Capillaries (from Lat. *capillus*, a hair) are the very minute blood-vessels which form the *terminations* of the smaller arteries, and the *commencement* of the smaller veins. It is impossible to draw the lines of demarcation, showing exactly where the arteries terminate, or the veins commence. (See fig. 47).

The *capillaries* in the body of a man are microscopic, cylindrical or sub-cylindrical tubes, having an average diameter of about $\frac{1}{3.000}$ of an inch; those in the brain being much smaller. They *differ* from the *veins* and *arteries*, not only in their greater *minuteness*, but also in the *structure of their walls*, which consist of an exceedingly fine homogeneous membrane containing *nuclei*, but destitute of *smooth muscular fibre*.

The *red corpuscles* are *too large* to admit of their passage into the smaller capillaries. The *capillaries* are the chief agents of *nutrition,* the *liquor sanguinis* or

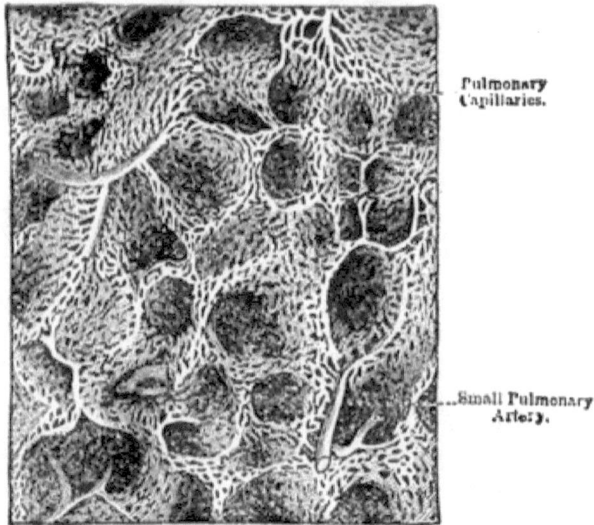

Fig. 47. Showing the Capillaries and small Arteries of the Air-
Cells of the Lungs. (Highly Magnified.)

blood-plasma exuding or passing by *exosmosis* through
their walls, and thus nourishing the adjacent tissues.

The *sectional area* of the capillaries has been estimated at about 400 times that of the arteries. It has
also been estimated that the blood moves 400 to 500
times more slowly in the *capillaries* than in the larger
arteries. In the capillaries it moves at the rate of
about 1½ inches per minute.

209. The Veins are the blood-vessels which *return* the
blood from the *capillaries* to the heart. They commence
in a large number of very minute vessels, continuous
with the capillaries, and, after uniting together as the
small twigs of a tree unite to form larger ones, ultimately form into two large *trunk-veins*—the *superior*
and *inferior venæ cavæ*—by which the blood is poured
back into the heart.

The *deep* veins in general accompany the arteries,
usually running side by side with them; these are

termed the *venæ comites* (companion veins), and bear
the same names as their companion arteries.

The *superficial* veins in general lie between the skin
and the outside of the muscles.

To stop a bleeding vein *pressure* should be applied to
the vein on the *opposite* side of the *wound* to which the
heart is situated.

210. The veins are more numerous, larger, and thinner
walled than the arteries ; they are consequently *flaccid*
and collapsed, when empty, as we see them in joints from
the butcher.   When a *vein* is wounded, the blood issues
from it in a *uniform continuous* stream, quite unlike the
jerking flow from the *arteries*.   The veins differ from the
arteries also in being abundantly supplied with *valves*,
which direct the blood towards the heart.

211. The Valves of the Veins consist in general of
two or three *pocket-shaped pouches*, or *semi-lunar folds* of
the membranes of the inner and middle coats of the
vein.   Immediately behind the point of connection of
each valve the vein is a little expanded to allow of the
blood's getting behind and closing the valve when any
attempt at *regurgitation* is made.

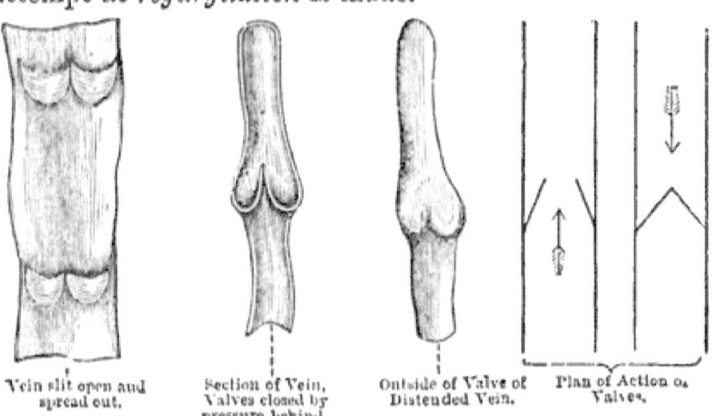

Vein slit open and   Section of Vein,   Outside of Valve of   Plan of Action of
spread out.   Valves closed by   Distended Vein.   Valves.
pressure behind.

Fig. 48. Showing semi-lunar Valves of Veins.

The *structure* and *action* of each valve are similar to
those of the aortic, pulmonary, and Eustachian valves
previously described.

212. Evidence of Circulation Obtainable in the Living Body.

(1.) When special poisons or soluble salts, as prussiate or nitrate of potash, are injected into the veins in one part of the system, they can readily be detected in blood drawn from *remote* veins in the course of a few seconds only.

(2.) Alcohol and other substances taken into the stomach can be detected in the blood and in the urine very shortly after its injection.

(3.) If a vein be wounded, pressure applied *between* it and the heart will not affect the bleeding; whereas pressure on the *remote* side of the wound will immediately arrest it.

(4.) On the contrary, if pressure be applied between a wound in an artery and the heart, it will *arrest* the bleeding; whereas if it be applied on the *opposite* (the *remote*) side of the wound, as in the case of the vein, the bleeding will *not* cease.

(5.) If pressure be applied to a vein *above* a valve, the progress of the blood is *arrested*, the vein *swells* up, and the exterior of the valve assumes a knotted appearance.

(6.) The movement of the corpuscles of the blood may be distinctly seen in the *transparent web* of a *frog's* foot, or the *tail* of a *tadpole* when placed under the microscope.

The most satisfactory evidence of the circulation, however, is that furnished by the *structure* of the *heart* and *blood-vessels,* and more especially by the arrangement of the *valves* in the veins, which is such as only to make it possible for the blood to move in one way *out* and one way *back* to the heart. This, however, can only be learnt, as Harvey, the great discoverer of the circulation learnt it—viz., by the study of the dead body.

CHAPTER VIII.

RESPIRATION.—THE LUNGS.

213. The Part Played by Oxygen in the Animal Economy.—The *vital power* by which the human body performs its various functions—physical, muscular, and mental—is derived from the *chemical force* eliminated during the combination of the *oxygen* of the air with the *carbon* and *hydrogen* of the *tissues;* just as in the case of the *steam engine,* the whole of the *work done* is executed by the *heat* developed during the *chemical* union of the

oxygen of the atmosphere with the *carbon* and *hydrogen* of the coal, or other fuel employed.   The presence of *oxygen* at every point of the body, where *work* is to be done, becomes thus an indispensable necessity.   This *vivifying* of the tissues, or supplying them with *oxygen*, and getting rid of the dead *products of their oxidation*, or combustion, is the chief function of *respiration* or *breathing*.

214. Respiration may be defined as the process by which oxygen is *absorbed* by the blood through the medium of the *lungs* and *skin;* while *carbonic acid*, *water*, and *urea*, are simultaneously *excreted*.

215. Changes in the Blood during Respiration.—When the dark purple *venous* blood, loaded with carbonic acid and other waste products, is brought by the pulmonary capillaries *to*, and distributed over the *surface* of the air-cells, the latter being simultaneously filled with air, the oxygen of the air being first *dissolved* in the liquid moistening the surface of the air-cells, passing by *endosmose* through their walls, and also through those of the capillaries, is immediately *absorbed* by the red cor-puscles of the blood, changing the blood to a *bright scarlet* colour.   *Simultaneously* with the absorption of the oxygen, an equal volume of *carbonic acid gas* is *dis-charged* by *exosmose* into the air-cells, to be expelled from the lungs by the process of *expiration*.

The *dark venous* unrespired blood thus differs from the bright *scarlet arterial* blood, the product of respira-tion, in containing not only *less* carbonic acid, urea, and water, but much *more* oxygen.

216. Proofs of Waste through the Lungs.—

(1.) Breathe through a glass vessel containing clear *lime water*.   The water immediately becomes *white* and *turbid*, from the formation of *chalk* or white carbonate of lime, thus proving carbonic acid gas to be evolved from the lungs.

(2. Breathe on to a *cold* bright looking-glass or bright steel knife blade, it immediately becomes *dim* from the deposit of *dew*, thus proving the evolu-tion of *water* from the lungs

(3.) Breathe through a small quantity of strong
sulphuric acid in a glass vessel, after a time it
blackens or turns brown, thus showing the
presence of *organic* matter in the breath.

This last experiment can only be performed without great or even
fatal danger by the practised chemist.

217. The Scheme of Structure of the Lungs consists
of (1.) A means of collecting and presenting, in the
smallest possible space, a very extensive *surface of blood*,
to a very large *surface of air;* of (2.) A means of collect-
and presenting, in an equally small compass, a large
surface of *air* to the *blood*.

A very good idea of the plan of structure of the lungs
may be formed by imagining a large sheet of fine silk,
containing about 1,400 square feet, to be packed up in
the chest, so as to leave myriads of minute *air cavities*
communicating with the external atmosphere. But in
this case each *minute* fibre of silk, of which the woven
tissue is composed, must be supposed to be not a *solid
fibre*, but a *hollow tube*, through which blood is circulating. In any case the student should, if possible,
endeavour to see a portion of *lung tissue*, especially its
capillary structure, under a good microscope. Without
such an opportunity, he can form no adequate idea of
the exquisite beauty of the minute lung structures.

218. The Lungs are the principal organs of respiration.
They consist of two large, light, conical, pinkish, mottled,
spongy, *elastic* organs, which surround the heart, and
occupy the right and left sides of the cavity of the *thorax*
or chest. They weigh about *twenty-four* ounces. The
right lung is larger, and consists of three *lobes* or divisions, separated by clefts; the *left* contains but two *lobes*.
These lobes again are divisible into *lobules*, each *lobule*
constituting, in fact, a *miniature lung.*(See figs. 41 & 49.)

Each lung consists essentially of an aggregation of air
tubes, air sacs with cells, arteries, veins, and capillaries,
with nerves and lymphatics.

When an infant has once *lived*—that is, has once had

its lungs inflated with air—they become lighter than, and will therefore float in, *water*. Hence, if a dead infant has been born alive, and a piece of its lung be cut off, it will *float* in water; if born dead, it will *sink*. As the *lungs* are the only viscera which will *float* in water, they are popularly termed, and are known by the butcher as "the *lights*."

EXPERIMENT.—Blow down the *trachea* into the "lights" (lungs) of a sheep; if not injured they will immediately become *distended*. Allow the air to escape and their *elasticity* will cause them to contract and *expel* the air.

219. Course of the Air in Breathing.—When the mouth is *closed*, the air on its way *to* the lungs passes through the two *anterior nares* (or nostrils) into the *nasal* cavities; from the nose through the two posterior nares into the *pharynx*; thence through the *larynx*, *trachea*, two *bronchi*, *bronchial tubes;* and ultimately, by slow mixture or *diffusion*, into the *air cells*. During its *return*, it of course takes the *opposite* direction. When the *mouth* is also open during breathing, a portion of the air passes through it and the *fauces* into the *pharynx*, where it joins the rest of the *tidal air* on its way to the lungs.

220. The Air Tubes of the Lungs consist of an *arborescent* system of tubes, comprising the *trachea*, right and left *bronchi*, primary, secondary, tertiary, and *ultimate* bronchial tubes, by which the air is distributed to all parts of the lungs. (See fig. 49.)

221. The Trachea (from Gr. *trachus*, rough), or windpipe, is the principal air tube of the lungs. It consists of a *membranous* tube, supported and kept open by sixteen to twenty *imperfect rings* (about two-thirds of the circle) of *cartilage* (gristle). It is about four and a-half inches long, and three-quarters of an inch in diameter. It *commences* at the bottom of the *larynx*, or voice box, and extends down to the lungs, opposite to the *third* dorsal vertebra, where it *bifurcates* or divides into two branches, termed respectively the right and the left *bronchus.* Its *interior* surface is lined with *ciliated mucous* membrane.

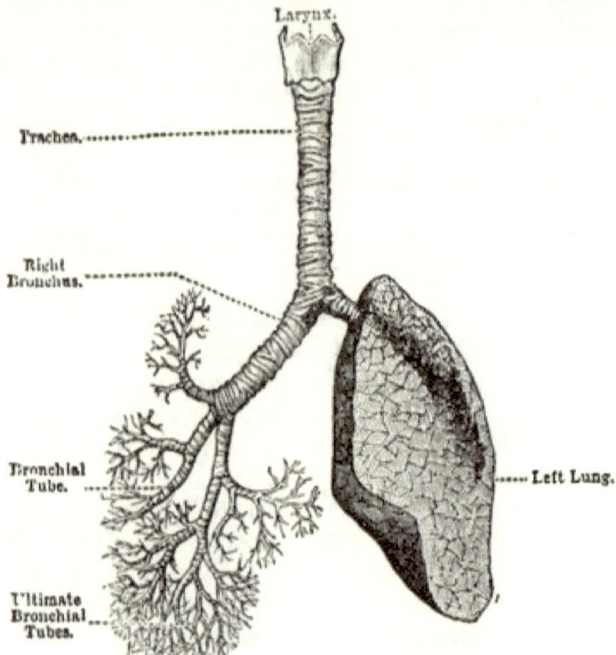

Fig. 49.—Left Lung and Air Tubes.

222. The Bronchi are the air tubes which enter the right and left lung. They are formed by the *bifurcation* of the *trachea,* and have the same general structure as that of the *trachea,* with the exception of their cartilaginous rings, which are *entire* rings, and not broken, as in the case of those of the trachea. (See fig. 49.)

223. The Bronchial Tubes are the smaller air tubes, formed by the *dichotomous* divisions and subdivisions of the two *bronchi,* which divide and subdivide like the branches of a tree. (See fig. 49.)

224. The Ultimate Bronchial Tubes are the smallest of the bronchial tubes; they terminate in the *infundibula,* and are formed by the last or *ultimate* division of these tubes. Their walls are exceedingly thin; they are lined with *tessellated* or *squamous* epithelium.   They contain no *cartilaginous* structures.

Inflammation of the bronchial tubes is termed *bronchitis.*

225. The Infundibula, or Lung Sacs, are minute funnel-shaped or conical expansions of the *terminal* or *ultimate* bronchial tubes; they are about *one-fortieth* of an inch in diameter.   Their walls are, as it were, punched out into very minute pouches, indentations, or *saccules,* termed

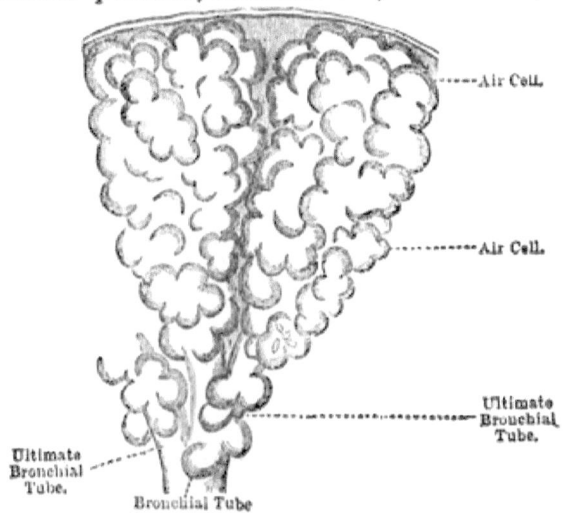

Fig. 50.   Showing Two Infundibula or Air-Sacs at the termination of Bronchial Tubes.

air cells.   (See fig. 50.)   These walls consist of exceedingly delicate, *fibrous* and *mucous* membrane, covered on the inside with *tessellated epithelium,* and lined on the outside by an exceedingly close *capillary* net-work, the diameter of the blood-vessels of which does not exceed the $\frac{1}{3,000}$ of an inch.   (See fig. 47.)   The passages or spaces in the lung sac between the air cells, and into which they open, are called *intercellular* passages.

226. The Air Cells, or Alveoli are the minute pouches or *saccules* on the walls of the *infundibula* just described. They are about $\frac{1}{700}$ of an inch in diameter, and it is said that there are as many as 1,700 in each *in-*

fundibulum, and that the lungs contain 600,000,000 of these *air cells.* (See fig. 50.)

227. The Chief Blood-vessels of the Lungs are the two pulmonary *arteries* which take the *venous* blood from the *right ventricle,* and distribute it to the capillaries of the lungs, and the four pulmonary veins which return the blood from the lungs to the *left auricle.* (See figs. 3, 41 and 43.)

228. Composition of Atmospheric Air.—Every 100 parts of ordinary unrespired air contains nearly 21 parts of *oxygen* gas, and nearly 79 parts of *nitrogen gas,* together with a minute portion of *aqueous vapour,* and a still more minute portion, about $\frac{1}{3,000}$, of *carbonic acid* gas.

The following Table shows more approximately the composition of *ordinary* and *respired* air, supposing (which is not the case) that the quantity of *aqueous vapour* remains unchanged.

ATMOSPHERIC AIR—PURE.	RESPIRED.
Oxygen, ... 20·61	16·26
Nitrogen, ... 77·95	77·95
Carbonic Acid, . . 0·04	4·39
Aqueous Vapour,. . 1·40	1·40
100·00	100·00

229. Changes in Respired Air.—Respired air, as it leaves the lungs, differs from ordinary air—

(1) In containing about four and a half to five per cent. *less* oxygen.

(2) In containing about four and a half to five per cent. more carbonic acid.

(3) In being *saturated* with aqueous vapour.

(4) In containing *urea* and other highly decomposable animal matter.

(5) In its comparatively high temperature, usually about 98° Fahrenheit.

The average quantity of solid carbon breathed out daily in the form of carbonic acid gas would be equal to a *lump of pure charcoal* weighing *seven* to *eight* ounces.

230. **Ventilation** (from Lat. *ventus*, the wind) is the process by which bad, vitiated, or *respired* air is systematically and continuously removed from a room or chamber, and its place supplied with pure, unrespired air.   No room is fit for human habitation which is not properly ventilated.   No human being should sleep in a room, or live, or work in a workshop, which is less than *nine* feet square, or which allows less cubical space per head than 700 to 800 cubic feet—viz., the cubical contents of such a room; and the whole of the air in such a room should be changed by a proper system of ventilation at least twice each hour.   The cubical space here mentioned is that professedly allowed in all well-appointed hospitals.

That an amount of pure air, equal to that just given, is required for healthy subsistence, is sufficiently proved by the following facts:—An average adult passes through his lungs per day nearly 400 cubic feet of air, which he robs of nearly five per cent., or nearly 20 cubic feet, of pure *life-giving* oxygen, and *poisons*, by discharging into it about the same bulk of carbonic acid gas and decomposable dead organic matter.   Suppose such a man to be hermetically closed up in such a room, in the course of twenty-four hours there would remain in that room no portion of the gas which had not passed into and out of his lungs, or which had not contributed its five per cent. of its oxygen towards the sustenance of his *vitality*, or which was not laden with five per cent. of his *burnt tissues*.

231. **Mechanical Movements of Respiration.** — The process of *respiration* is effected by means of the alternate *enlargement* and *diminution* of the cavity of the chest or *thorax*, the movements in this process very nearly resembling the action of an ordinary *bellows* in blowing a fire.

When the cavity of the chest is *enlarged*, air rushes in, impelled by its own *pressure* or *elasticity:* this constitutes *inspiration*. When the cavity of the chest is *diminished* by the pressure and weight of its own walls, a portion of the enclosed air is forced out : this constitutes *expiration*.

The chief causes of these movements are :—1*st*, The *mobility* of the walls of the chest, the chief agents of which are the respiratory muscles ; 2*nd*, The *elasticity* of the lungs.

232. **The Thorax** or chest is a closed, air-tight, conical, or *bee-hive* shaped cavity, containing but one opening. It is situated in the upper half of the *trunk*, its base being turned downwards so as to rest on the top of the *abdomen*.

The *interior* cavity of the thorax is occupied by the *heart, lungs*, and great blood-vessels. (See fig. 3.) The heart, the lungs, and the walls of the chest are lined by *serous* membranes, termed respectively the *pericardium* and the *pleuræ*.

The *walls* of the thorax are built up of the twelve pairs of *ribs* or *costæ*, the *costal cartilages*, the *sternum*, the twelve *dorsal vertebræ*, and the *intercostal* muscles. The *thorax* is closed in below by a floor of *tendon* and *muscle*, termed the *diaphragm*. (See fig. 13, and Appendix fig. 83.)

The *bony* and *cartilaginous structures* of the chest form a sort of *cage*, the openings or interstices of which are filled up by muscles.

233. **The Diaphragm or Midriff** is the large thin *musculo-tendinous*, somewhat fan-shaped septum, partition, or dividing membrane, which separates the cavities of the thorax and the abdomen, and forms the *floor* of the thoracic, and the *roof* of the abdominal cavity. (Fig.51.)

234. **Ordinary Inspiration**. — During inspiration the cavity of the chest is enlarging—*vertically*, by the contraction and consequent descent of the diaphragm ; *laterally* and *antero-posteriorly*, by the *twisting* and elevation and consequent *lifting* or pushing *outwards*

of the ribs and sternum, and consequently of the walls of the chest.

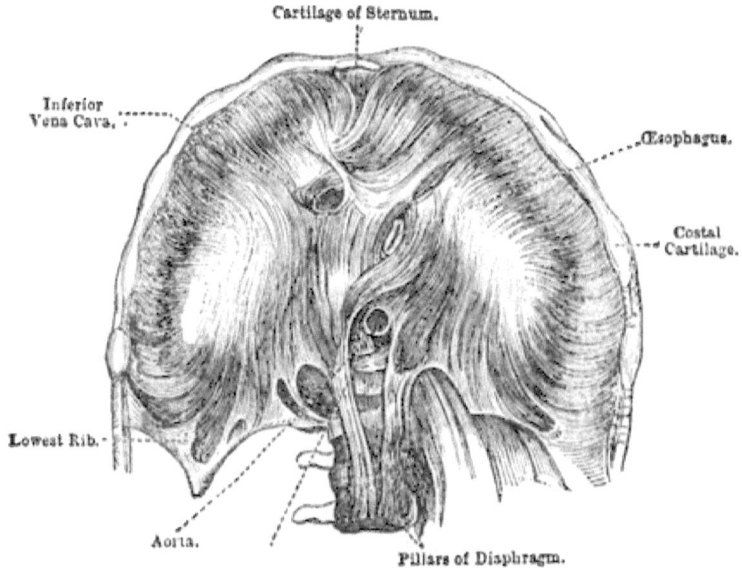

Fig. 51. Showing Diaphragm seen from the lower or Abdominal Side.

As the cavity of the chest is thus enlarging, a tendency to the formation of a *vacuum* is produced. This is, however, simultaneously counteracted by the pressure of the external atmosphere, which, pressing with a force of fifteen pounds on the square inch, immediately rushes down the *air-tract* (that is, through the *mouth*, or *nose*, or both) into the *pharynx*, thence successively through the *glottis* into the *larynx*, the *trachea*, the *bronchi*, the bronchial tubes, and the air sacs and cells of the lungs.

The various movements here indicated are produced as follows :—

(1) The *levatores costarum* (muscles which elevate the upper ribs) contract, slightly raise, and *fix* the upper

14 E   H

ribs, which thus become a base of operations for the
lower ribs. (See fig. 83, Appendix.)

(2) The *external intercostal* muscles, which fill up the
spaces between the ribs ; also, a portion of the internal
intercostals and of the trianguli sterni now contract,
drawing *up* and *outwards* the remaining ribs.

(3) The *diaphragm* (a large flat muscle, fixed by its
outer rim or periphery) now contracts from its tendinous
centre, thus causing it to descend.

The various movements here described really take
place simultaneously. There are also other muscles
than those now mentioned, which take part in these
changes, which have been omitted for the sake of greater
simplicity.

During very gentle respiration this work is effected
almost entirely by the movements of the *diaphragm.*

235. Ordinary Expiration is chiefly effected by the
elastic recoil of the ribs, cartilages, and the lung tissues,
and by the action of *gravity*, which causes the ribs and
other organs to fall into their original positions so soon
as they are released from the action of the *inspiratory*
muscles. Expiration is, in fact, mainly due to the return
of the lungs, ribs, and diaphragm, to the condition they
were in *before* inspiration. The *internal intercostals*,
muscles which pull the ribs downwards, also probably
aid in even *ordinary* expiration.

236. Frequency of Respiration.—An ordinary adult at
rest breathes on the average *eighteen* times per minute;
during disease the number of respirations may amount
to 100 per minute. The number of respirations per
minute increases very greatly with exertion, as during
running. It also increases with the *rarity* of the atmosphere, being greater up high mountains than on the
plains below.

237. Asphyxia or Suffocation is the term applied to
that condition of " *oxygen starvation* " which is induced
by the prolonged suspension of the process of respiration. If sufficiently prolonged, it terminates in death.
When death is caused by *suffocation*, it arises from the

cessation of the action of the heart and the consequent *arrest of the circulation.*

CHAPTER IX.

ANIMAL HEAT.

238. Animal Heat is, as has been shown, a consequence of *respiration.* It is *generated* wherever the blood or tissues of the organism are *oxidated* or converted into carbonic acid, water, urea, or uric acid. Every capillary vessel, every point in the tissues external to the capillaries in which this act of combustion takes place, becomes practically a " small fire-place, in which heat is being evolved in proportion to the activity of the *chemical* changes which are going on."

239. The *Animal heat* is *regulated* by the skin and by the organs of circulation. The former keeps down the temperature through the agency of the *perspiration,* by which *evaporation* is promoted in a ratio proportionate to the surplus heat generated—the latter by distributing it uniformly through the body.

Nothing tends to keep down the accumulation of heat more than *evaporation,* by which heat is rendered *latent* or insensible.

It is thus that the skin regulates the animal heat in a Turkish bath, or that a *live* animal might *live* in a *hot* oven in which a *dead* one would be *roasted.* In other words, the *perspiration* keeps down the animal heat. (See *Sudoriparous Glands.*)

240. The Average Temperature of the interior of the human body is 98° to 100° Fahrenheit; but it varies slightly with age, health, exercise, climate, &c. In dangerous fevers, it rises 8° or 10° above this, in consequence of the skin's not performing its functions properly. One of the first objects of the physician, therefore, is to restore the healthy action of this organ.

Clothing promotes the animal heat only by lessening the rapidity of its escape.

CHAPTER X.

241. Digestion (from Lat. *dis*, asunder; and *gestus*, carried), in its larger sense, is the process by which the *nutritious* are separated from the *innutritious* or useless parts of the food, and converted into blood.

As these processes for the most part take place in the *alimentary canal*, they are frequently described as constituting the function of *alimentation*.

242. General View of the Alimentary Canal.—The alimentary canal is a *musculo-membranous* tube, about 26 feet long, which, commencing at the *mouth*, passes, by a series of coils or *convolutions*, the whole length of the body, and terminates at the *rectum*. It consists of *four* coats or layers. Commencing at the *mouth* and the *lips*, and continued into the *pharynx*, it afterwards, in its downward course, forms the *œsophagus* or gullet; then, *expanding largely*, it forms the *stomach;* again contracting, it forms the *small intestines;* and lastly, again *expanding*, it forms the *large intestines*.

243. General View of the Course of the Food and the Changes it undergoes.—The solid food, on entering the mouth, is *masticated* or broken up into minute portions and mixed with the saliva (spittle), *swallowed*, and mixed with an *acid* juice poured into it on its entrance into the stomach, the liquid (the *gastric juice*) flowing out of its walls. Here the *proteids* are more or less *dissolved* and *absorbed* by the *veins* of the stomach; also so much of the *starch* of the food as has been under the influence of the *saliva* converted into *sugar*. When thus prepared in the stomach it is called *chyme;* the *chyme* then passes into the *small intestines*, where it is mixed with two other juices, this time *not acid*, but *alkaline*, poured out to meet it and make its *fatty* portions soluble, or at least *absorbable*. It is now termed

chyle. It is gradually *worked* or pressed along the
intestines, all the *useful* or *nutritious* parts being gra-

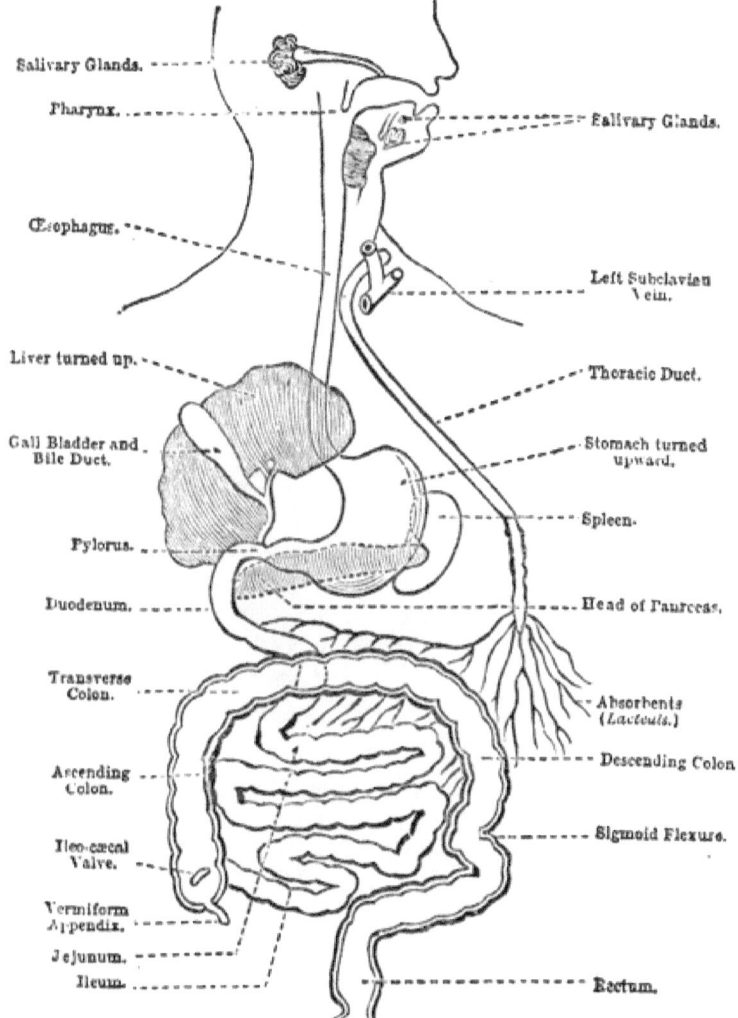

Salivary Glands.

Pharynx.

Œsophagus.

Liver turned up.

Gall Bladder and
Bile Duct.

Pylorus.

Duodenum.

Transverse
Colon.

Ascending
Colon.

Ileo-cæcal
Valve.

Vermiform
Appendix.

Jejunum.

Ileum.

Salivary Glands.

Left Subclavian
Vein.

Thoracic Duct.

Stomach turned
upward.

Spleen.

Head of Pancreas.

Absorbents
(*Lacteals.*)

Descending Colon

Sigmoid Flexure.

Rectum.

Fig. 52. Showing Course of Food.

dually *absorbed* by the *lacteals* and *veins*, until its arrival
at the end of the *large intestines,* where it is expelled
from the body.

244. The *fatty* parts *absorbed* by the *lacteals* make their way directly by the *thoracic* duct to the *left subclavian vein*, and thence by the *upper vena cava* to the heart. The *dissolved proteids* and the *sugar* (and probably some of the *fats*) are, however, obliged to take a *longer* and more *circuitous* route on their way to the heart. They are first collected from the *veins* of the *stomach and intestines*, then passed by the *portal vein* to, and circulated through the liver; after which they are passed with the blood from the liver into the *hepatic* veins, hence to the lower *vena cava*, and by it poured into the heart.

245. The Mouth is the *irregular*, somewhat oval-shaped cavity which forms the commencement of the alimentary canal, and in which the food is *masticated*.

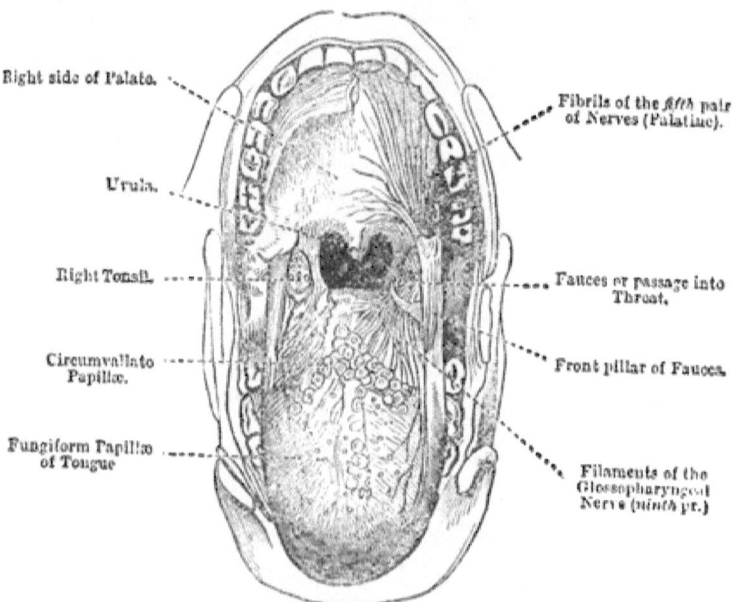

Fig. 53. The Mouth showing Tongue, Palate, &c.

It is bounded in front by the lips; on the side, by the cheeks and portions of the *upper* and *lower* jaws; above, by the *hard palate* which forms its roof; behind, by the *soft palate* and *fauces;* and below, by the tongue and mucous membrane, reaching from beneath it to the front of the inside of the lower jaw, which form a sort of movable floor. (See fig. 53 and 55.)

The cavity of the mouth is separated from that of the nose by the *hard* and *soft* palates ; the latter, also, with its prolongation, the *uvula*, and with the *epiglottis*, separate it from the *pharynx*.

When the mouth is shut the *dorsum*, or back of the tongue, touches the palate.

246. The mouth contains the tongue and the teeth, by which, with the aid of the jaws and the salivary glands, the food is *masticated* and *insalivated*. It and the tongue are lined with *mucous* membrane, which is more or less studded with little *buccal* glands, about the size of millet seeds, which supply through their ducts, which open into the mouth, a portion of the necessary moisture. The *parotid, sub-maxillary*, and the *sub-lingual* salivary glands, however, supply nearly all the saliva that enters the mouth.

247. Mastication, or Chewing, is the process by which the food is broken, crunched, and ground by the teeth, aided by the tongue, cheeks, lips, and jaws. The object of the process is to overcome the *force of cohesion*, and thus promote the *solution* or liquefaction of the food. It is effected by an up-and-down or *vertical*, an *antero-posterior* or front to back, and a *lateral* or side-to-side (the two latter together producing an *oblique*) motion of the jaws.

248. Teeth.—Man is provided during life with two *sets* of teeth—the *first* of which appear during infancy—termed the *temporary, deciduous* or *milk teeth;* the *second* set, which *begins* to appear during childhood, but which is not *completed* until the *wisdom*-teeth have appeared, about the commencement of *adult* life, are termed the *permanent teeth.*

249. The Permanent Teeth, when complete, are 32 in number: they are arranged in the form of *arches* in the sockets or *alveoli* of the upper and lower jaws.

Each dental arch (gum) contains 16 teeth. These

sixteen teeth contain *four* types, shapes, or varieties, as shown in the diagram (fig. 54)—viz., four *incisor*, two *canine*, four *bicuspid*, and six *molar* teeth.

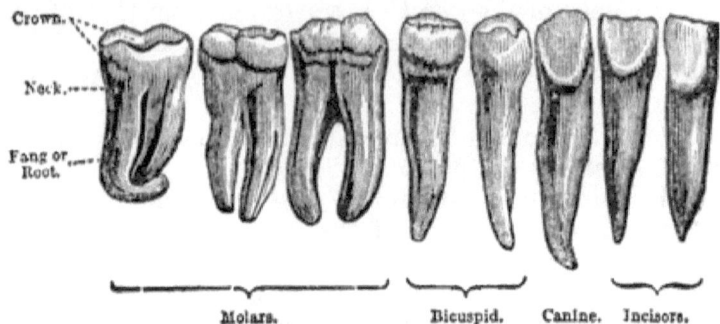

Fig. 54.—Showing Differently Shaped Teeth.

250. Insalivation is the process by which the food during mastication is mixed with *air* and *saliva*. The *saliva* facilitates swallowing by *lubricating* the food,—it makes the *starch* in the food *soluble*, by ultimately converting it into *sugar*. It also makes the food more permeable to the juices of the stomach. It promotes *taste* by dissolving the sapid substances in the food.

251. The Salivary Glands are the three pairs of *conglomerate* glands which secrete the saliva. The largest are the *parotid glands*, situated immediately below and in front of the ear: they weigh from ½ oz. to 1 oz. each. Their *ducts*, about 2½ inches long, open upon the inner surface of the cheek by orifices opposite the upper second molar teeth.

The *sub-maxillary glands* are situated near the neck, in the lower jaw, under the floor of the mouth; their *ducts*, about 2 inches long, open under the tip of the tongue.

The *sublingual glands* are also situated, as their name implies, under the tongue; but not quite so far back as the *sub-maxillary*. Their *ducts* also pour the saliva into the mouth under the tip of the tongue.

252. The Saliva or *spittle* is the thin, watery, slightly viscid and frothy liquid poured into the mouth from the *buccal* and *salivary glands.* It usually contains a little *mucus*, also *epithelial scales*, which render it slightly *opalescent.*

It contains a small quantity of a peculiar *nitrogenous* principle, capable of converting starch into sugar; this principle, which does *not* act on *fats* or *albuminous* substances (proteids), is termed *ptyalin.*

253. The Pharynx (from Gr. *pharugx*, the gullet) is the funnel-shaped part of the alimentary canal, which is placed immediately behind the mouth, nose, and larynx. It is separated from the cavity of the mouth by the *soft palate, uvula,* and the *epiglottis* (see fig. 55). It is about

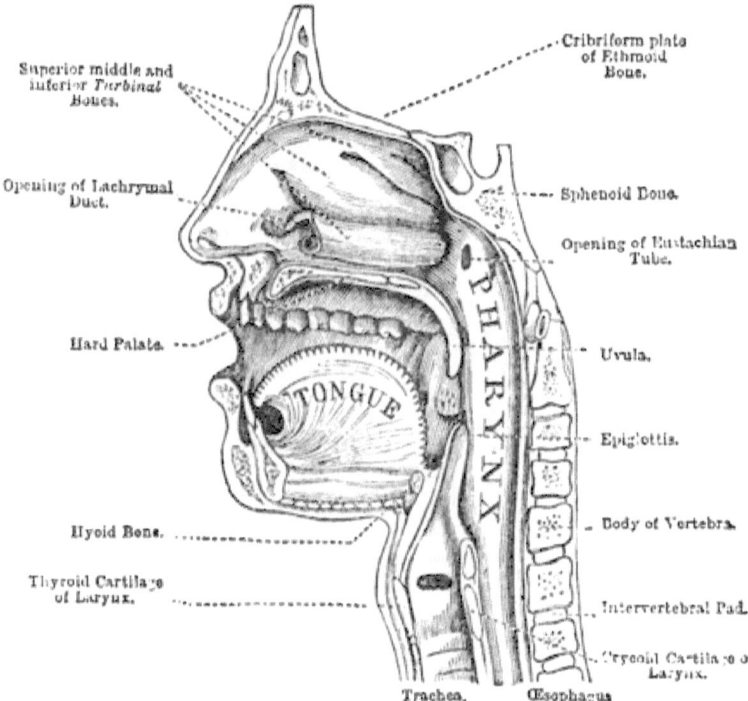

Fig. 55. Section of Mouth, Nose, Pharynx, Œsophagus, Larynx, Trachea, &c.

4½ inches long, varies from 2 inches at its upper to 1

inch at its lower extremity, and has seven openings into it—viz., the two *posterior nares* (nostrils), the two *Eustachian tubes*, the *mouth, larynx,* and *œsophagus.* Like the rest of the alimentary canal, its interior is lined by *mucous* membrane. It is supplied with several muscles.

254. **The Œsophagus** (from Gr. *oisō*, I carry; and *phago*, I eat), gullet or food-pipe, is the *musculo-membranous* tube which forms that part of the alimentary canal which passes from the *pharynx* through the *diaphragm* to the *cardiac orifice* of the stomach. It is about 9 inches long, and forms the narrowest part of the alimentary canal. Its *upper* part is supplied with *striated* muscular fibre, its lower part with *non-striated* muscular fibre only. (See figs. 52 and 55.)

255. **Deglutition,** or *swallowing,* is the process by which food or drink is *forced* down the *œsophagus* or gullet into the stomach. That both solid food and drink do not simply *fall* by the action of *gravity* into the stomach is proved by acrobats and others, who sometimes perform the feat of eating and drinking " while standing on their heads." In the case of horses and other similar animals, the food and drink have to pass from the mouth upwards, against gravity, before they can reach the stomach.

256. **The Stomach** or principal organ of digestion is a large (when distended), bent, conical, or "bag-pipe" shaped bag, pouch, or expansion of the alimentary canal, capable of containing three to five pints of liquid. When moderately full, it is about 12 inches long, and about 4 inches in its larger diameter.

Its *left* extremity, the convexity of which lies against the concave side of the *spleen,* is termed the *greater* or *splenic* end: it contains the cardiac pouch, or dilatation; also termed the greater *cul-de-sac* or *fundus* of the stomach. Its *right* extremity, which is much smaller than the *left,* terminates at the *pylorus,* where it joins the *duodenum* or first portion of the small intestines.

257. The stomach has two orifices, the *œsophageal* or

cardiac orifice on the top, so called because it is on the same side of the body as the heart; and the right *pyloric orifice,* by which it opens into the intestines. The muscular

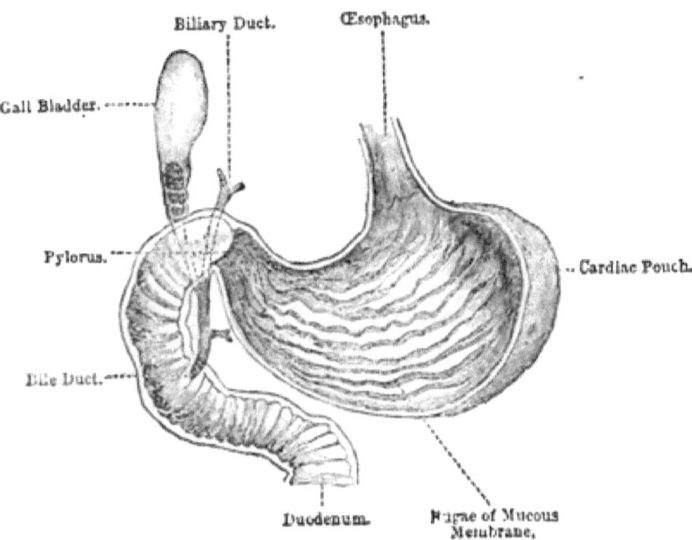

Biliary Duct.   Œsophagus.

Gall Bladder.

Pylorus.

.. Cardiac Pouch.

Bile Duct.

Duodenum.   Rugae of Mucous Membrane.

Fig. 56. Showing the Stomach and its interior lining of Mucous Membrane, the Duodenum with the Valvulæ Conniventes in its interior; also the Gall Bladder and Bile Ducts.

fibre around the latter is thickened and so arranged as to form a kind of *sphincter* muscle, termed the *pylorus* or *pyloric valve.*

The *walls* of the stomach consist of four coats, the inner or *mucous* coat of which is very complex, being abundantly supplied with minute follicles, the *gastric follicles* or *tubuli.*

258. Immediately food is passed into the stomach, it begins to *contract,* and *roll* the food about by its *peristaltic* action, while simultaneously the gastric juice is poured out of the numerous follicles in its walls, and thus becomes thoroughly mixed with it. These movements were actually observed by Dr. Beaumont in the stomach of a patient who had suffered from a gun-shot wound.

The stomach lies *transversely* across the upper part of the front of the cavity of the abdomen (see fig. 2), its *left* lying under the ribs and diaphragm, and in contact with the spleen; its *right* end underlies the liver.

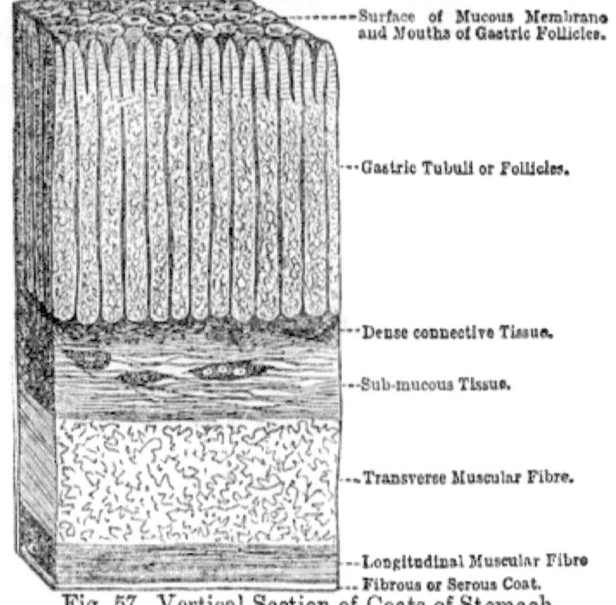

Surface of Mucous Membrane and Mouths of Gastric Follicles.

Gastric Tubuli or Follicles.

Dense connective Tissue.

Sub-mucous Tissue.

Transverse Muscular Fibre.

Longitudinal Muscular Fibre
Fibrous or Serous Coat.

Fig. 57. Vertical Section of Coats of Stomach.

259. The Coats of the Stomach.—The stomach is usually described as consisting of four coats—

1. An outer *fibrous* coat, the *serous* coat.
2. A *muscular* coat consisting of longitudinal, transverse, and oblique smooth (unstriated) muscular fibre, by which the peristaltic movement is carried on.
3. A *submucous* coat of *loose connective* tissue, forming a matrix, in which the blood-vessels and nerves break up and ramify before reaching the mucous coat.
4. A *mucous* coat, consisting of a layer of *basement membrane*, covered by an inner layer of *epithelial cells*. Its surface is covered with minute *shallow* pits or *alveoli*. The bottoms of these shallow pits are studded with the mouths of the *gastric follicles* which dip into this membrane. This membrane is abundantly supplied with nerves and blood-vessels; its surface is greater than that of its other coats: it therefore collects in *rugae* (folds) when the stomach is empty, as shown in fig. 57.

260. The Gastric Juice is the clear, colourless, or pale straw coloured, slightly acid liquid secreted by the *mucous membrane* of the stomach and its follicles. It readily mixes with water, has powerful *antiseptic* properties, *coagulates* albumen, and at 90° to 100° Fahr. is a good solvent of *proteid* or *albuminoid* substances, converting them into a liquid termed *peptone*. It not only does not appear to act on *fats* and *starches*, but arrests the action of other juices upon them.

Its *solvent* power over the *proteids* appears to depend upon the presence of a peculiar principle termed *pepsin*. It also contains *hydrochloric* or *lactic* acid. The following table shows its approximate composition:—

Gastric Juice,
- Water, about . . 99·2
- Pepsin, ,, . . . ·3
- Hydrochloric acid, ·2
- Salts, about . . . ·2

261. Chymification, or **Gastric Digestion,** is the process by which the food is converted into *chyme*, by the action of the *saliva* and *gastric juice*.

262. The Chyme is the slightly *acid* gruel or pea-soup like more or less viscid product of *gastric* digestion; it varies considerably in appearance and consistence with the nature of the diet. It consists of a heterogeneous *mixture* of various substances, comprising chiefly the *indigestible* portions of the food, the *amyloids* (starchy substances), not yet converted into *sugar*, and the sugar and *peptone* not yet absorbed; also, more or less *saliva* and *gastric juice*.

263. The Pylorus (from Gr. *pule*, gate; *ouros*, guardian), or pyloric valve, is a sort of *sphincter*, or *ring*-shaped muscle, formed by the reduplication of *mucous* and *muscular* membranes of the stomach. It encircles and regulates the size of the *pyloric* aperture, only allowing, until the stomach becomes exhausted from over-work, the *finer* portions of the *chyme* to *strain* through into the duodenum. (See figs. 52 and 55).

264. Large and Small Intestines.—The *small intes-*

tines, consisting of the *duodenum, jejunum,* and *ileum,* are *coiled* up in the abdomen, being enclosed in the *mesentery,* and attached by it to the spine.   They form a tube about 20 feet long.   The surface of their interior lining of mucous membrane is increased by the *valvulæ conniventes,* which consist of small transverse circular folds of *mucous* membrane, which promote absorption by the gentle resistance they offer to the passage of the food.

The *large intestines* commence at the end of the *ileum,*

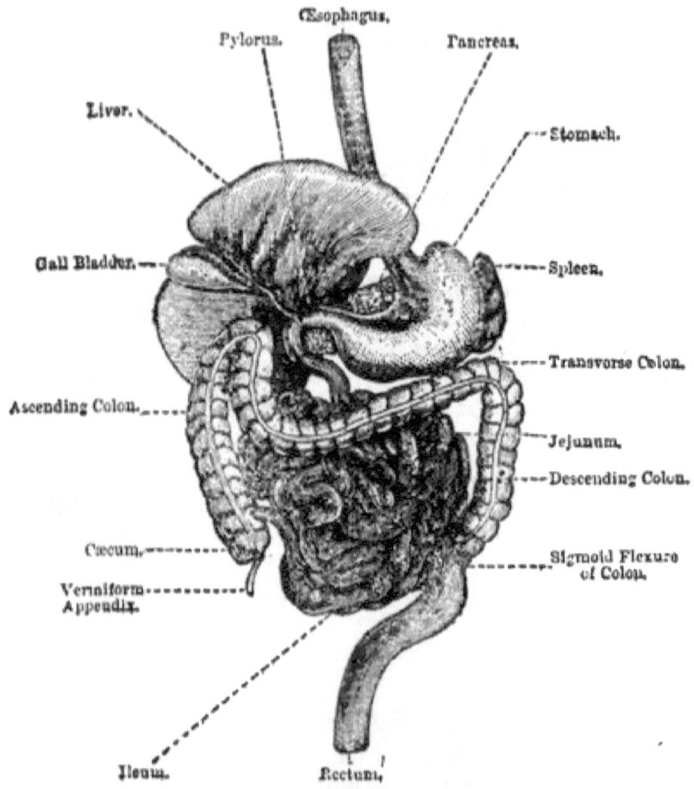

Fig. 58. Large and Small Intestines.

and terminate at the *rectum.*   Their exterior surface, excepting that of the rectum, differs from that of the small intestines in being "puckered up."   This puckered appearance is due to three *flat bands of muscular fibres* which run along their exterior, and which, being *shorter* than the tube to which they are attached, produce the puckering referred to. They are about six feet long, and consist of the *cæcum*, the *colon* (ascending, transverse, and descending the sigmoid flexure), and the *rectum*.

The *mucous membrane* lining the interior of both the large and the small intestines is more or less covered with *villi*, *Peyer's glands*, and *Lieberkühn's* follicles.

265. Intestinal Digestion.—Chylification.—The *chyme*, having passed into the small intestines, receives and mixes with the *bile* and the *pancreatic juice*, by which the *fatty* parts of the food are reduced to the form of a sort of *emulsion*, termed *chyle*, which is absorbed by the *lymphatics* of the intestines (the *lacteals*). The conversion of the starch into sugar, and its absorption, as also that of the remaining *peptone* by the veins, are likewise completed in the intestine. The *indigestible* parts of the food are pushed along, losing more and more of the nutritious matter, the peptone, fats, and sugar mixed with it, until, after passing through the *cæcum*, they become more or less acid, acquire the peculiar offensive fæcal odour, and ultimately passing through the *rectum*, where they collect until they are expelled as fæces. (See fig. 52.)

CHAPTER XI.

FOOD AND NUTRITION.

266. Food or aliment may be defined as consisting of those *external nutritious* substances which are passed into the *alimentary canal* for the purpose of being *digested*, and restoring the *losses* of the system.

267. Classification of Food.—Food is sometimes classified under *two* heads:—

(1.) *Heat-forming*, fuel, or respiratory food, which does contain *carbon* and *hydrogen*, which serve as sources of heat, but which does not contain *nitrogen*, and, therefore, which cannot completely restore the *lost* or *wasted* tissues.

(2.) *Flesh-forming*, plastic, albuminoid, proteid, or *nitrogenous* food, which *does* contain *nitrogen*, in addition to *carbon, hydrogen,* and *oxygen*, and is therefore capable of *completely* restoring the *lost* tissues.

268. Food Stuffs may be divided into *four* classes, as shown in the following Table. For a *description* of the respective substances contained in the Table, see sections given :—

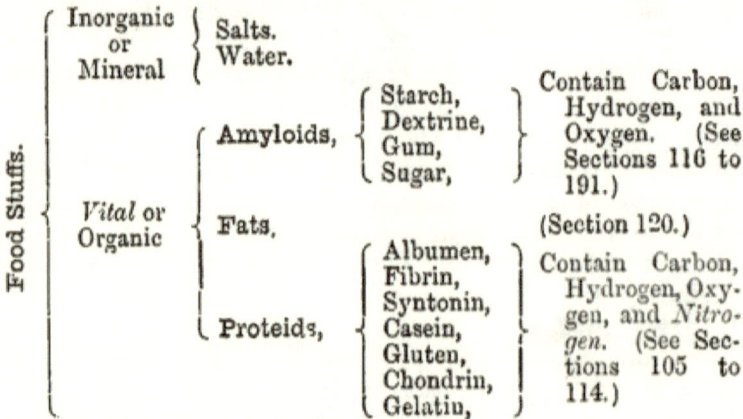

269. Economical Admixture of Food.—To obtain the requisite 4,000 grains of carbon per day from dry *proteids*, a man would be compelled to eat about 7,500 grains; but this would give him nearly 1,200 grains of *nitrogen*, or nearly *four* times the quantity of nitrogen he would require. That is, if he lived on *fatless* meat, he would require 5 to 6 ℔s per day to give him the necessary carbon; whereas he might get the necessary *carbon* and *nitrogen* from 4 to 5 ℔s of *bread*, or from a mixed diet, consisting of 2 ℔s of bread and ¾ ℔ of meat. He might also get the same from about 1 ℔ of *fatless* meat, and ½ ℔ of *fat*, or 1 ℔ of *sugar*. If he attempted to get the necessary quantity of *nitrogen* from a purely potato diet—a comparatively *innutritious* diet—he would probably be compelled, in order to get the necessary *nitrogen*, to eat 10 ℔s to 12 ℔s or upwards per day. On the other hand, if he tried to live on a highly *nutritious* and exclusively *proteid* diet, he might die of starvation in consequence of the great loss of vital power he would sustain in the digestive attempts to get the *necessary carbon*, under these unfavourable conditions. In fact, in all probability, after a time, his vital powers would give way in his attempt to obtain the materials necessary for subsistence.

270. **Nutrition** (from Lat. *nutrio*, I nourish) is the process by which the tissues of the living body are repaired, built up, or their loss of substance restored out of material supplied them by the *liquor sanguinis* or *blood plasma*. The chief agents of nutrition are the capillaries.

CHAPTER XII.

THE LYMPHATICS, LACTEALS, AND THORACIC DUCT.

271. **The Lymphatic or Absorbent System** consists of the *thoracic duct*, right lymphatic duct, receptaculum chyli, the lymphatic vessels, the *lacteals*, and the *lymphatic* and *mesenteric* glands (see fig. 59,) which shows the most important of these organs as they appear injected with mercury.

272. **The Thoracic Duct** is the terminal and largest or main trunk of the lymphatic system. It is about 18 or 20 inches long, and about the diameter of a moderate sized goose-quill. Commencing in the abdomen, somewhere opposite the *second lumbar* vertebra, it passes through the diaphragm by the aortic opening, ascending near the vertebral column to the neck; it then curves downward, and joins the *subclavian* vein at the angle formed by its junction with the *jugular* vein. Its structure is very similar to that of the veins, and it is numerously supplied with *valves* (see fig. 59). The *right lymphatic duct* is small and unimportant; it contains no *chyle*.

273. **The Lacteals** are the *lymphatic* vessels of the
14 E.   I

intestines. They are so called because of their *milky*
appearance, when filled with *chyle, two* or *three* hours
after a meal. The primary lacteals commence in the

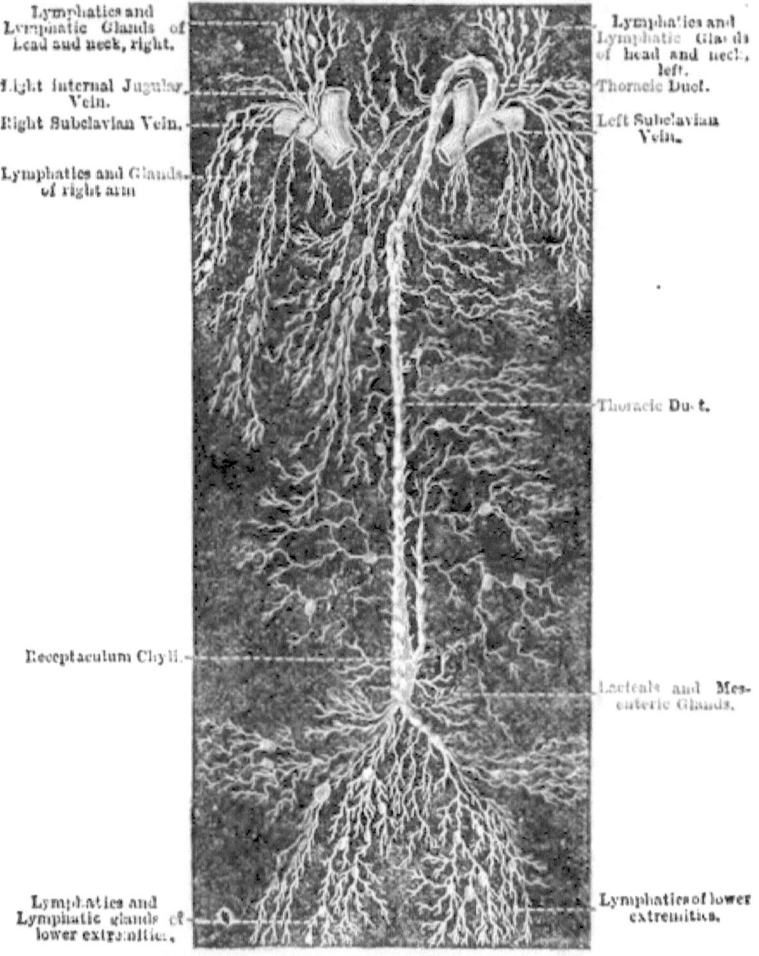

Lymphatics and
Lymphatic Glands of
head and neck, right.

Right Internal Jugular
Vein.

Right Subclavian Vein.

Lymphatics and Glands
of right arm

Lymphatics and
Lymphatic Glands
of head and neck,
left.
Thoracic Duct.

Left Subclavian
Vein.

Thoracic Duct.

Lacteals and Mes-
enteric Glands.

Receptaculum Chyli.

Lymphatics and
Lymphatic glands of
lower extremities.

Lymphatics of lower
extremities.

Fig. 59.   Showing the Principal Lymphatics of the Human Body.
The small nodular bodies are the *Lymphatic Glands.*

middle of the *villi* of the intestines, where they form
microscopic club-like tubes, or minute microscopic
plexuses. (See fig. 59.)

274. The Lymphatics are the vessels which *absorb* and elaborate the *lymph*, and convey it to the *thoracic* duct. For a long time the smaller lymphatics escaped detection, because of the *transparency* of their walls and of the contained *lymph*. There are two sets of lymphatics —the deep and the superficial. They are distributed to all parts of the body, except the brain, the spinal cord, and the interior of the eye. (See fig. 59).

275. The Lymphatic Glands are small, solid, rounded or oval, pinkish, glandular bodies, which lie in the course of the lymphatics, and through which they pass on their way to the *thoracic duct* (see fig. 59.) They vary in size from that of a hemp-seed to that of an almond. The lymphatic vessel which passes *to* the gland is termed the *afferent* vessel; that which *leaves* it, the *efferent* vessel. The chief lymphatic glands in the body are—the *cervical* (neck), the *axillary* (arm-pit), the *lumbar* (loins), the *inguinal* (groin), *mesenteric* and the *femoral* lymphatics (inside the thigh). The *lymph* in the *efferent* vessels contains a greater number of *lymph corpuscles* than that in the *afferent* lymphatic vessels.

276. The Mesenteric Glands are the *lymphatic* glands of the *lacteal*, so called because they are contained within the folds of the *mesentery:* they assist in elaborating the chyle.

277. The Lymph is the transparent, colourless, *coagulable* liquid contained in the lymphatics. It is supposed to consist of those portions of the *liquor sanguinis* which, having *exuded* through the walls of the capillaries, and bathed the tissues for the *purpose* of their *nutrition*, has *not* been *appropriated* by them and has therefore been *absorbed* by the *lymphatics*, to be restored by them and *re-mixed* with the blood. It resembles diluted *liquor sanguinis*, but contains a few *colourless corpuscles* resembling those of the blood.

Lymph differs from *blood* in containing more water, fewer *white*, and no *red* corpuscles.

278. The Chyle (from Gr. *chulos*, juice) is the *milky*-white, *coagulable*, more or less *fatty* and *alkaline* liquid which is formed in the intestines after the mixture of the *bile* and *pancreatic* juice with the *chyme*. It contains minute colourless corpuscles and oil globules. It is chiefly absorbed by the *lacteals* during the passage of the food through the *intestines*, and conveyed by them through the *mesenteric* glands, where it becomes more or less organized by the development of *chyle corpuscles* (which resemble, though smaller, the *colourless corpuscles* of the blood) to the *thoracic duct*, by which it is poured into the *left subclavian vein*, and thus mixed with the blood.

The *fatty* portions of the food find their way into the blood as *chyle*.

CHAPTER XIII.

SECRETION AND EXCRETION.

279. Secretion is the process by which solids or liquids, differing from the constituents of the blood, and necessary for the proper performance of the functions of the body, are elaborated or separated from it by means of *glands* or other organs, as in the secretion of the *saliva*, and t he *gastic* and *panceatic* juices.

280. Excretion is the process by which waste, useless and injurious matter, is separated from the blood and thrown out of the body by the *excretory glands*.

281. A Gland is an organ whose function is that of *secretion* or *excretion*, or both combined, and which contains *ducts* or vessels for the escape of matter elaborated or excreted by the gland. Secreting organs which do not contain ducts are not termed glands, as the mucous and serous membranes. The principal glands of the body are the liver, the kidneys, the pancreas, the mammary, and the lachrymal glands. Other less important glands are the sudoriparous glands, the sebaceous, the ceruminous, the Meibomian glands, and the glands of Brunner, Peyer, and Lieberkühn.

282. The Liver is the *largest* gland in the body; it is incessantly in action, secreting *bile* and *glycogen*, and is therefore a *constant* source of *gain* and *loss* to the blood. (See figs. 2 and 58.)

It is the large *reddish-brown* organ situated at the right side of the top of the abdomen, immediately underneath the right belt of the diaphragm, to the concavity of which it is attached, its left lobe overlying a portion of the stomach.

The liver is usually described as being 3 to 5 lbs in weight, and as *secreting* 3 to 5 lbs of *bile* per day. It is 10 to 13 inches long, 6 to 7 inches broad, and about 3 inches at its greatest thickness.

283. Structure of the Liver.—The liver contains five *lobes* and five *fissures*, and is supported in its situation by five *ligaments*, chiefly folds in the *peritoneum*. The two chief lobes are the *right* and *left* lobes, the former being six times as large as the latter.

The liver is almost entirely enfolded by the *peritoneum*, which forms its *serous* coat; but besides this it has its own peculiar *fibrous coat* or tunic, which passes by the *portal canal* into the substance of the liver, with the blood-vessels and hepatic ducts investing them, and thus forming the sheath termed " *Glisson's Capsule.*"

284. The Lobes of the liver consist of an agglomeration or collection of *lobules.*

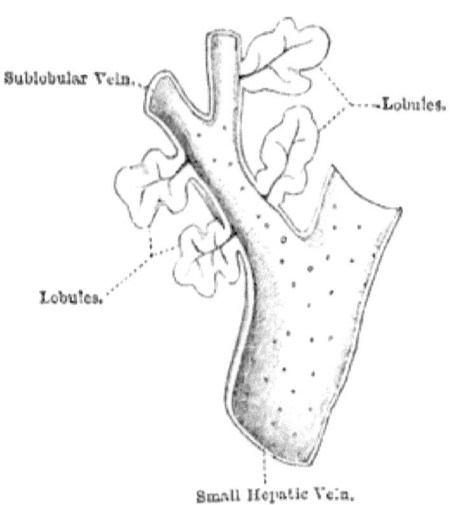

Fig. 60. Diagram of Lobules of Liver.

285. The Lobules are small, roundish or *granular* bodies, about $\frac{1}{20}$ to $\frac{1}{10}$ of an inch in diameter, or about the size of a pin's head or of a *millet* seed. They consist of masses or agglomerations of *biliary* or *hepatic* cells, which are clustered about the minute branches of the *portal* veins, which form the *interlobular* veins. (See figs. 60 and 61.)

286. **The Biliary Ducts** are the minute ducts or tubules by which the bile is collected from the cells, and ultimately conducted to the gall bladder or the intestines. Their mode of origin is still undetermined.

287. **The Cystic Duct,** which is about an inch long, conveys the bile, secreted during the intervals when digestion is not proceeding, into the *gall bladder*.

288. **The Hepatic Duct** is formed by *two* trunks, which

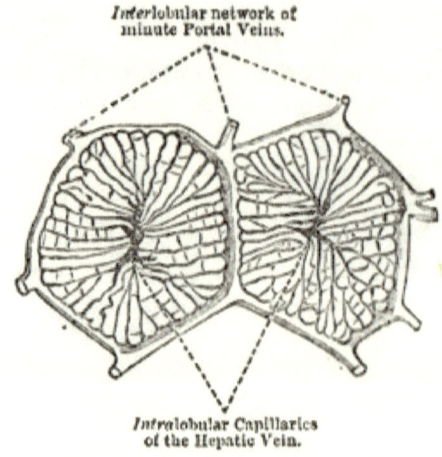

Interlobular network of minute Portal Veins.

Intralobular Capillaries of the Hepatic Vein.

Fig. 61. Transverse Section of Two Lobules of the Liver, showing minute branches of the Portal Veins (Interlobular), encircling Biliary Lobules.

pass out of the right and left lobes of the liver, and unite.

289. **The Common Bile Duct** (*ductus communis choledochus*), which is about the diameter of an ordinary goose quill, and about three inches long, is the largest of the bile ducts. It is formed by the junction of the *cystic* and *hepatic* ducts, and passes *obliquely* on its termination

between the *muscular* and *mucous* coats of the duodenum, entering that organ, with the *common* opening of the *pancreatic duct*, at about its middle. The perforated *walls* of the duodenum thus act as a valve, permitting or obstructing the flow of the bile as the *duodenum* is *full* or *empty*. When the duodenum is *active*, and its walls well charged with blood, they are pushed out, the canal becomes relieved, and the *bile* readily forces its way through the distended aperture; but when it is empty and inactive, its walls *collapse*, and the sides of the *duct* become folded together, and the passage thus stopped. These *ducts* consist of an *external* fibrous and an *internal* mucous coat.

290. **The Gall Bladder** is the *pear-shaped*, conical, membranous bag, which is seated in a cavity under the *right* lobe of the liver. It is about 4 inches long, and 1 inch in its greatest diameter, and holds 8 to 10 drachms of bile. It consists of an outer *serous*, a middle *muscular*, and an inner *mucous* coat. Its function is to serve as a *reservoir* for the bile, during the *intervals* between digestion, when the aid of the bile is not required. (See figs. 56 and 58.)

291. **The Portal Vein,** which carries and distributes to the liver the blood laden with the *products of digestion* absorbed through the walls of the veins of the stomach and intestines, is formed by the *union* of the veins from the stomach, intestines, pancreas, and spleen. The *portal vein* passes into the liver through the transverse or *portal fissure*, and then subdividing, after the *manner of an artery*, sends out minute branches, which, passing along the *portal canals*, ultimately encircle the lobules, thus becoming *inter*lobular veins, and then give off minute capillaries, the *portal capillaries*, which, passing into the substance of the lobules—that is, between the walls of the *biliary cells*—supply them with the blood from which the *bile* is *secreted*.

292. **The Hepatic Veins** are the veins by which the *used-up* portal and arterial blood of the liver—that is, the blood supplied to the liver for the *formation of bile*,

and also that supplied to *nourish* and *vivify* the liver itself—is collected and poured into the *vena cava*. The *minute* hepatic veins or capillaries commence in the centres of the lobules, thus forming the *intra*-lobular vessels.

293. The Hepatic Artery is one of the branches of the *cœliac axis*. It supplies the blood which nourishes the membranes, the coats of the large vessels, and the ducts of the liver.

294. The Bile is a yellow, or greenish-yellow, viscid, extremely bitter, slightly odorous and slightly *alkaline* fluid. It is a little heavier than water, its sp. gr. being about 1020. Its chief *function* is apparently to aid the digestion of *fatty* matters, by *neutralizing* the *acid* of the gastric juice and converting the fat into an *emulsion*.

The *bile* consists of *water*, *mucus*, and from 10 to 17 per cent. of *solid* matter.

The *solid* matter consists chiefly of *bilin;* but it also contains fat, *cholesterine*, and salts.

The *bilin* is a *resinous* substance, composed of *glycocholic* and *taurocholic* acids, in combination with *soda*.

The *bile* consists of carbon, hydrogen, oxygen, nitrogen, and sulphur; the carbon and hydrogen being in great *excess*, as compared with that of the blood; the nitrogen of the bile being, on the contrary, greatly *deficient*, as compared with that of the blood.

The secretion of the bile may thus be regarded as an *excrementitious* process in relation to the elements *carbon* and *hydrogen;* also as a *digestive* process in relation to *fatty* substances.

295. Glycogen (Gr. *glukus*, sweet; *gennao*, I produce) is a peculiar whitish, tasteless, soluble amyloid substance, possessing properties intermediate to those of *hydrated starch* and *dextrin*, which is *formed* in the liver. It is distinguished by its strong tendency to change into sugar in the presence of any animal *ferment*. This substance may be readily detected in the blood of the *hepatic vein*, and in the *ascending vena cava*, just before it enters the heart; but it cannot be detected in blood which has passed through the lungs. Evidently, therefore, it has disappeared by combustion or oxidation, —thus playing the part of respiratory fuel, or heat food.

296. Proof of the Glycogenic Function of the Liver.— If blood be drawn from the *hepatic artery* or the *portal vein*—that is, *before* or *as* it *enters* the liver—no *glycogen* can be detected in it. But if blood be drawn from the hepatic *veins*—that is, as it *leaves* the liver—*glycogen* may be detected in abundance; thus proving its formation in the liver itself. Further, if the liver be repeatedly washed out by the frequent injection of water into its vessels until all traces of *glycogen* disappear, after allowing the liver to remain at rest for some time, *glycogen* will however again manifest itself in its substance.

297. Glucose, or liver sugar, is the saccharine substance formed by the action of animal ferments on the glycogen. It disappears from the blood after it has passed through the lungs.

298. The Pancreas, or sweetbread, is the long, soft, hammer or tongue-shaped, milky-white gland which lies under the back of the stomach, extending from the spleen to the duodenum (see fig. 58). Its *tail*, or smaller end, is in contact with the spleen; its larger end, or *head*, lies in the bend or concavity of the duodenum. It is six to seven inches long, and weighs three to five ounces, and probably secretes about half a pint of *pancreatic juice* per day.

The *pancreas* has been termed the abdominal salivary gland.

299. The Pancreatic Juice is a colourless, nearly tasteless, slightly viscid, *alkaline* fluid, in general appearance and properties very similar to the saliva. It acts powerfully on *starchy* substances, and aids the digestion of the *fats*, but is not supposed to act on the *proteids*.

CHAPTER XIV.

EXCRETION—THE SKIN

300. The Skin, or outer integument (Lat. *tego*, I cover) of the body, is the somewhat complex organ, both in structure and function, which *invests* and surrounds the exterior of the body. It, with some modification of structure, passes as *mucous* membrane by the mouth into, and lines the interior of the *alimentary canal* which passes through the trunk.

301. Functions of the Skin.—The chief functions of the skin are the following:—

(1.) It *regulates* the *temperature* of the body, or the *animal heat.*

(2.) It *protects* from air, dirt, and injury, and binds together the superficial organs of the body.

(3.) It is an organ of *excretion* and *absorption.*

(4.) It is an organ of *sensation* or *touch.*

(5.) It is an organ of *respiration*, it absorbs *oxygen*, and evolves *carbonic acid.* (This perhaps is properly included under 3.)

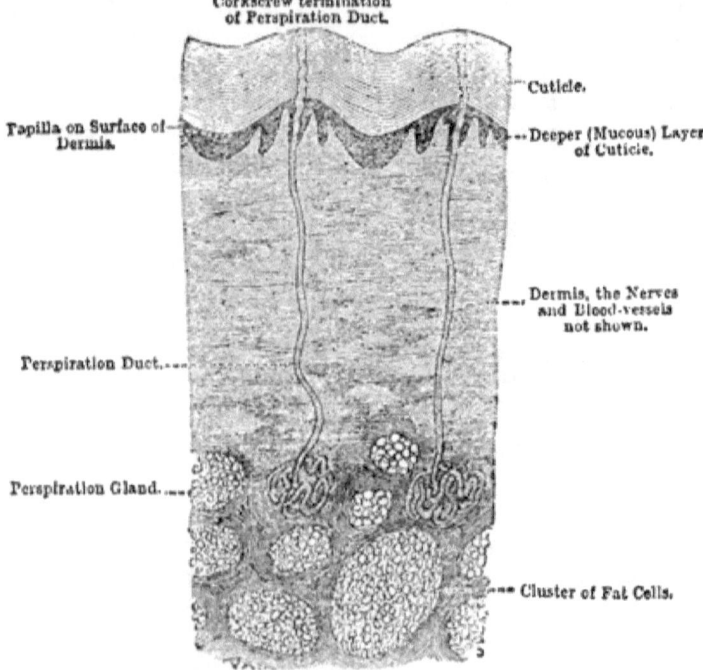

Fig. 62. Vertical Section of Skin.

302. Structure of the Skin.—The skin consists essentially of *two* distinct layers or formations, viz.:—(1) the cuticle; and (2) the cutis.

303. (1.) The Cuticle is the outer *non-vascular* layer of *epithelial cells* which lines and protects the cutis. It is variously called the *cortical layer* of the skin, the *epidermis* or the *scarf skin*. The *inner* or lowermost layer of the epithelial cells, of which it is composed, is moulded on to the true skin or dermis. The cells of this, the softer and lowermost layer, are more round and moist than those of the outer layer; and, in the case of the negro and other coloured races, contain dark *pigment* or colouring matter. This *layer* is sometimes called the *mucous layer* of the cuticle, or the *rete mucosum.*

The *cuticle* is the layer of the skin which is raised up from the lower one by the collection of clear fluid, when a blister is formed. It is quite devoid of sensibility when pricked.

304. (2.) The Cutis, variously termed the *dermis, corium, cutis vera* or *true skin,* forms the more important and complex lower or under layer of the skin. It consists of—

(*a*) A fibrous network or matrix of connective tissue.

(*b*) A network of blood-vessels (capillaries) ramifying in the meshes of the former.

(*c*) A network of nervous fibrils, also ramifying through the same.

(*d*) *Sudoriparous* and *sebaceous glands* and ducts, and masses of adipose tissue (fat cells), enclosed within the fibrous matrix.

The *fibrous matrix,* when tanned, forms leather, and when boiled, yields *gelatine* (jelly). The closeness of the nervous fibrils and of the capillary blood-vessels is shown by the fact, that nowhere in the substance of the true skin can the point of a needle be inserted without causing *bleeding* and *pain.*

305. Sudoriparous Glands (Lat. *sudo,* I sweat).—The sweat, perspiration, or sudoriparous glands which *excrete* the perspiration, consist of minute *coiled tubules,* formed of *basement membrane,* lined with *epithelial* cells. (See fig. 62.)

306. The Cutaneous Perspiration, or the sweat, consists, when condensed, of a colourless, transparent, slightly *acid* liquid, having a peculiar and characteristic odour. It contains carbonic acid, *urea*, and lactic, also traces of formic, acetic, and butyric acids. It also usually contains small quantities of *sebaceous* matter and of *epithelial cells*, which render it more or less turbid. It is constantly given off from the skin in the form of *invisible vapour*, or, as it is termed, *insensible* perspiration. But when its escape is prevented, or when it is given off very rapidly—as during great exertion, during violent mental emotion, or when the body is heavily clothed—it collects in the skin in the form of a *liquid*, and is then termed *sensible* perspiration.

The skin is intermediate in its function to that of the *lungs* and the *kidneys*. It, like the lungs, *respires*—that is, absorbs oxygen and excretes carbonic acid; but it also, like the kidneys, excretes *urea*. The skin is more active in *hot*, the kidneys in *cold* weather.

307. The Sebaceous Glands (Lat. *sebum*, suet) consist of minute *sacculated bags* or *follicles* of *basement membrane*, lined with *epithelium*, and containing more or less of the *fatty* matter they *secrete*. They are most numerous about the *hair follicles*, which are usually supplied with a pair of glands, and in the substance of those parts of the skin which undergo much flexion. Their function is apparently to keep the skin soft and flexible.

308. The Papillæ of the Skin consist of little conical processes on the surface of the *cutis* or true skin, immediately below the epidermis. Their extremities are sometimes simple, sometimes divided (see fig. 62). They are most abundant on the fingers, the palms of the hands, and the soles of the feet. The central portion of each papilla contains a minute *plexus* of blood-vessels, comprising an *arterial* and *venous* loop, also a *nerve fibril*, though the latter is occasionally absent. On many of the more sensitive parts of the skin, as the lips and the palm of the hands, the papillæ are supplied with *touch corpuscles*.

The papillæ are about $\frac{1}{100}$ of an inch long, and $\frac{1}{250}$ of an inch in diameter at the base.

309. Compass Test of Sensibility.—The *sensibility* of the various parts of the skin, the epidermal covering being of equal thickness, is in general proportionate to the supply of *nervous fibrils* to these parts, and to the *activity* of their *circulation*.

310. A pair of drawing compasses, with their points slightly blunted, forms an excellent test (Weber's method) of the *relative sensibility* of the surface of the skin at different parts of the body, the *sensibility* of that part being greatest at which the two points of the compasses can be separately distinguished from each other when placed nearest together. The following Table shows the *relative sensibility* of different parts of the skin, as determined by this method :—

TABLE SHOWING SENSIBILITY OF VARIOUS PARTS OF SKIN
BY COMPASS-TEST.

Point of Tongue, -	½ line	Palm of Hand, -	5 lines
Tip of Finger, -	1 ,,	Forehead, - -	10 ,,
Red Surface of Lips,	2 lines	Back of Hand, -	14 ,,
Tip of Nose, -	3 ,,	,, Thigh, -	30 ,,

The distances given in the above Table are the shortest distances at which the two compass points give *two* distinct impressions; beyond those distances the two points give the impression of but one. The student should repeat these observations on himself.

CHAPTER XV.

THE KIDNEYS AND URINARY ORGANS.

311. The Chief Urinary Organs are the kidneys, ureters, and the bladder by which the urine is *excreted, conveyed* to the bladder, and *stored* up until its accumulation produces a sensation of uneasiness which ultimately leads to its expulsion. (See fig. 63.)

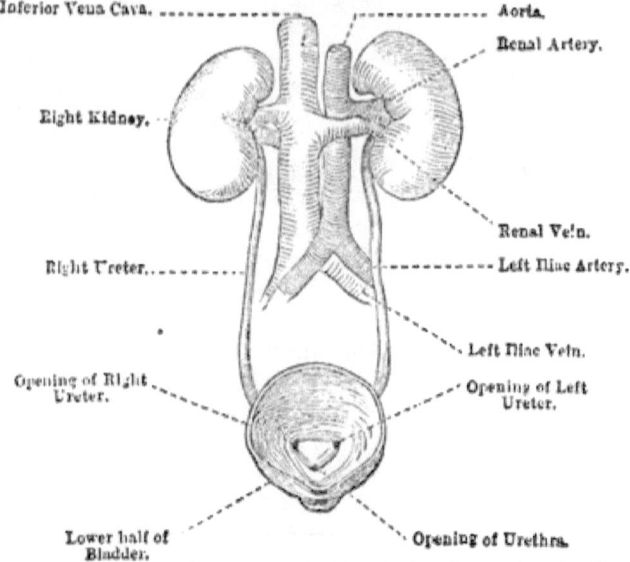

Fig. 63. Showing the Kidneys and their Blood-vessels, the Bladder and Ureters.

312. The Kidneys are the two bent, oval-shaped, dark-reddish organs situated outside the *peritoneum*, at the back of the abdominal cavity, one on each side of the spinal column, opposite the junction of the *dorsal* and *lumbar* vertebræ.

The bent or internal borders of the kidneys, which are turned a little backwards and towards the spine, contain openings termed *hiluses,* which give *entrance* to the *renal* arteries and nerves, and *exit* to the *renal* veins and *ducts.*

The human kidney very closely resembles that of the sheep with which we are all familiar, but is a little larger, measuring about 4 inches long, 2 inches wide, and 1 inch thick. Each kidney is surrounded and protected by a large mass of loose *connective* tissue and fat—as seen in a loin of sheep at the butchers.

313. Cut a sheep's kidney through the middle, *longitudinally,* from its *convex* to its *concave* border, with a very sharp knife, and the following structures, passing from without inwards, will be observable:—

(1.) An *external* thin membranous covering of connective tissue

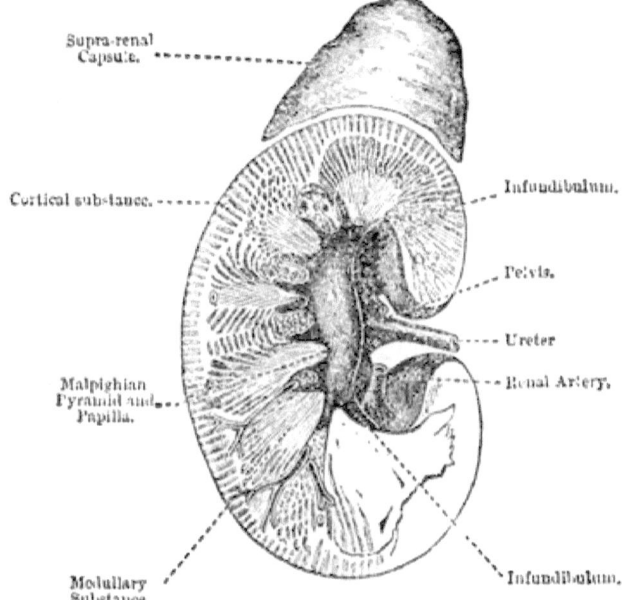

Fig. 64. Longitudinal Section of Human Kidney and Suprarenal Capsule.

—the *capsule*—which passes in at the *hilus* and lines its interior *sinus* or cavity.

(2.) A darker, somewhat spotted, friable layer, about one-fifth of an inch thick, termed the *cortical* substance, which forms about three-fourths of the substance of the gland.

(3.) A *fibrous* striated-looking structure, termed the *medullary* substance, the larger portion of which is collected into the form of pyramids (the Malpighian pyramids), the summits of which, termed *papillæ* or *mamillæ*, protrude into the *sinus* or cavity of the kidney.

(4.) A hollow cavity or *sinus* (dividing into *two* central and a terminal funnel-shaped portion, termed *infundibula*) which opens by the *pelvis* into the *ureter*. This cavity is lined by the *capsular* membrane. Those portions of the membrane which line the papillæ are termed the *calyces*.

314. The Medullary Substance of the kidney consists

of *straight tubules* of transparent *basement* membrane, varying from $\frac{1}{800}$ to $\frac{1}{200}$ of an inch in diameter, *lined* internally with *spheroidal* or glandular *epithelial cells*. Each *tubule* is surrounded by a minute *plexus* of veins, from which it derives the *urea* secreted by the *glandular epithelium*. About 1,000 of these *tubuli uriniferi* open and discharge their excretion (urine) into the sinus from the end of each pyramid.

315. **The Cortical Substance** of the kidneys consist chiefly of the *Malpighian capsules*, and their contained *glomeruli* or arterial tufts of the *convoluted* and *tortuous* continuations of the *straight tubuli uriniferi*.

The ends of the *straight tubules* first become *convoluted*, and then terminate into flask-like *dilatations*, the Malpighian capsules. Into this *capsule* an artery, termed the *efferent* artery, *enters* and coils or *loops* up into a little ball or tuft, termed a *glomerulus, pushing*

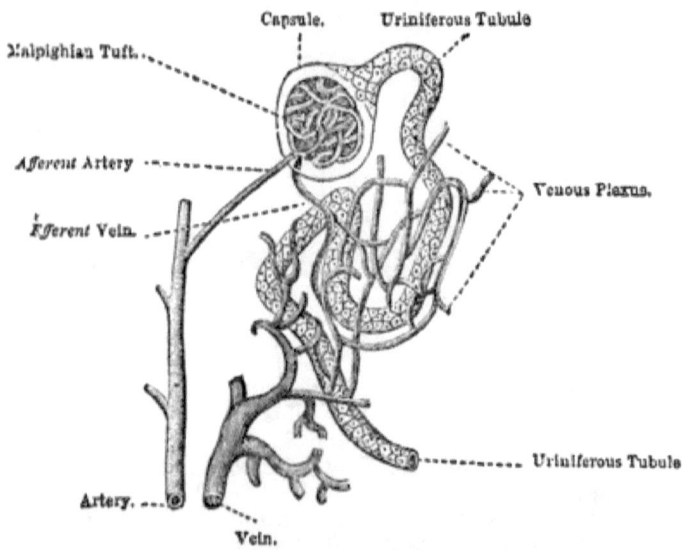

Fig. 65. Plan of Circulation in Kidney.

before it the membranous end of the capsule, so that the membrane completely invests the arterial tuft; the other end of the coiled blood-vessel leaves or passes out of the capsule as the *efferent vein* (see fig. 65). Professor Huxley compares this with the ordinary filtering arrangement of the chemist—the wall of the capsule representing the funnel, the walls of the arterial tuft the filtering paper, and the blood in its interior the liquid to be filtered—a portion of which, as it were, filters through into the tubule.

316. **Circulation of the Blood in the Kidneys.**—Fig. 65 shows the minute circulation in the kidneys. The arterial blood enters the kidneys by the *renal arteries,* which, after dividing and subdividing, ultimately distribute it to the *efferent* arteries of the various *glomeruli* or arterial tufts. The blood leaves each *tuft* by the *afferent* veins, which *first* form *venous plexuses* (resembling in miniature those of the *portal* circulation of the liver), which surround the *uriniferous tubules* and supply them with the blood from which the *urea* is secreted, and *then* unite to form the *radicles* of the *renal veins.*

317. **The Purest Blood** in the body is that passed from the kidneys by the renal veins, the urine excreted having removed not only most of its *nitrogenous waste* matter, but probably as much carbonic acid as it could have acquired during its short journey from the lungs. Therefore the blood will now contain its *minimum* quantity both of *carbonic acid* and *urea*; in other words, it will consist of the *aortic* blood deprived of most of its *nitrogenous waste matter.*

318. **The Ureters** are the *two excretory* ducts, about 16 or 18 inches long, and about the diameter of a goose-quill, which convey the urine from the kidneys to the bladder.

319. **The Bladder,** which is placed in the *pelvis,* serves as a *reservoir* to retain the continuously secreted urine, consists, when distended, of an oval-shaped *sac* or bag, about 5 inches long and 3 inches wide, and is capable of holding a pint and upwards of fluid.

The neck of the bladder is furnished with a *sphincter* muscle, by whose contraction the external *passage from* the bladder (the urethra) is closed.

14 E.   K

320. The Urine is the (when healthy) clear, pale-yellow coloured, acid, fermentible liquid secreted by the kidneys. It contains the chief *nitrogenous* waste of the system. One thousand parts of it contain about 965 of *water*, 14 of *urea*, ·4 of *uric acid*, 10 of extractive colouring matter, &c., about 10 of salts (chiefly sodium, calcium, and magnesium phosphates and sulphates), and common salt (sodium chloride).

The *urine* also contains a small quantity of *carbonic* acid, and still smaller quantities of *nitrogen* and *oxygen*. The urine thus contains the elements of the blood and tissues in a condition of disintegration.

321. The Spleen, or milt, is the very distensible, flattened, oval, dark reddish body, whose slightly concave inner surface lies upon the left side of the stomach. It probably assists in the formation of the *white corpuscles* of the blood.

CHAPTER XVI.

ANIMAL MECHANICS—THE MUSCLES, TENDONS, JOINTS, LIGAMENTS, AND LEVERS.

322. Animal Mechanics is that branch of Physiology which treats of the various movements of the animal body, or of motion and locomotion, and of the contrivances by which they are effected and the mode in which they are used. The muscles are the chief *active* agents of motion; the bones, ligaments, joints, and tendons are the *passive* agents.

A very fine delicate *microscopic* movement, which takes place over a considerable area of *membranous* surface in the interior of the body, is effected by means of *cilia*. The origin of this movement, which is effected by the *cilia* bending alternately *backward* and *forward* at their *base*, and which is entirely independent of the nervous system, is not understood.

323. The Muscles, which consist of red fleshy masses of *contractile* fibre, comprise *two* kinds, viz. :—(1.) *Hollow muscles*, which enclose cavities, and the contraction and extension of which alternately contract and expand these cavities, as explained in the case of the heart and the *alimentary canal;* (2.) *Solid muscles* which are for the most part attached to *bony levers*. *Hollow muscles* consist of *unstriped* or *smooth* muscular fibre; *solid muscles* consist of *striated* muscular fibre. (See secs. 142, 143). When a muscle *contracts,* its thickness *increases* in the same degree that its length decreases, so that it does not alter in actual *bulk*.

324. The Solid Muscles usually consist of masses of *contractile* fibre which are arranged, one end of the muscle being attached to a bony *lever* (that is, of a movable bone) the other end of the muscle being attached to a second (*fixed*) bone, a *joint* intervening between the ends of the two bones, so that when the muscle *contracts, one* of the bones is *moved* towards the other.

Such muscles usually possess a *belly*, or fuller and thicker, more or less *convex* mass in the middle, and two smaller *extremities* terminating in *tendon*, termed respectively the *origin* and *insertion* of the muscle.

The *end* of the muscle attached to the fixed bone is termed its *origin;* the *end* attached to the movable bone is termed the *insertion* of the muscle.

325. Tendons or Sinews (from Lat. *tendo,* I stretch) consist of the tough, flexible, but inelastic whitish *cords* or bands of *fibrous* tissue by which the muscles are attached to the bones, as the *tendo Achillis*, by which the *gastrocnemius* muscle (muscle of the calf of the leg) is attached to the *os calcis*. A good illustration of a *tendon* is presented in the yellowish-white cord in the leg of a fowl, which, as most boys know when pulled, draws up or closes the foot and claws.

The *tendons* play the same parts to the bones that the harness plays to the carriage. When the muscles contract they pull the tendons which pull the bones, just as when the horses pull the harness the harness pulls the carriage.

326. Ligaments.—See sec. 87.

327. The Articular Cartilages consist of the thin layers of true cartilage which tip the *surfaces* of the ends of the movable bones.   Its *smoothness* lessens *friction*, its *elasticity* lessens the concussion of the bones.

328. The Synovial Sacs (from Gr. *sun*, with ; and *ōon*, an egg) are kinds of *sacs* or bags which line the cartilages of the joints—thus forming a *double* layer of membrane, one layer adhering to one bone, the other to the other bone, as usual with the *serous* membranes, the interior of the sac, where the two substances rub together, being *lubricated* by a transparent *yellowish-white*, or reddish, glairy, viscid fluid secretion, in appearance resembling the white of an egg, termed the *synovia*.

The passages through which the *tendons* glide are also lined by *synovial* sacs or *bursæ*.

329. A Joint or Articulation consists of the union of two or more bones.   Movable joints consist of *perfect* and *imperfect* joints.

(1.) The *imperfect* joints are such as those of the *vertebræ* of the spine, which have no smooth linings of cartilage, and no synovial sacs, and which possess but very limited degrees of motion.

(2.) In the perfect joints the *ends* of the movable bones, at the surfaces of the joints, are tipped with *cartilage*, lined by *synovial sacs*, and *lubricated* with *synovia* in order to prevent friction.   They are held together by means of *ligaments*.

The principal joints of this class are the *ball and socket*, the *hinge* and the *pivot* joints.

330. The Ball and Socket Joints consist, like those of the arm and thigh, of rounded *heads*, fitting into rounded *cavities* or *sockets*.   They admit of very considerable motion in almost every direction, allowing the arms and legs to be *rotated* so as to describe a cone round an imaginary axis.   This movement is termed *circuminduction*.   When the socket is *shallow*, like the *glenoid cavity* of the shoulder, there is great freedom of motion ; but the bones are easily dislocated.

When the socket is deep, like the *acetabulum* of the pelvis, there is much less freedom of motion, but also *less chance of dislocation.*

331. The Hinge Joints, so called from their resemblance in plan of structure to a *common hinge*, only permit two motions — a *backward* and a *forward* motion in *one plane*. *Hinge* joints are single and double.

The *elbow* is a *single hinge* joint; the lower end of the *humerus* presents a nearly cylindrical head, which fits into a corresponding cavity in the *ulna*. The knee and the ankle are examples of less perfect single hinge joints.

332. In *double-hinge* or saddle-shaped joints, the end of *each* bone is *convex* from side to side in one direction, and *concave* from side to side in a direction at *right* angles to the convexity. They bear a certain limited resemblance to a *saddle* whose upper surface is *concave* from front to back, and *convex* from side to side. "A man seated in a saddle" is "articulated" with the saddle by such a joint. The *metacarpal* bone of the *thumb* is articulated to the *trapezium* (one of the carpal bones) by a *double-hinge* joint.

The joints of the *phalanges* of the hand and foot are essentially *hinge* joints.

333. Pivot Joints are formed by projections, processes, or pivots, on one bone, on to which, by a suitable ring or fitting, a second bone *fits* and *turns*, or in which the first bone *turns* on its *own axis*. Such joints permit of partial *rotation*. It is evident the *rotation* could not be *complete* without causing the laceration and destruction of the neighbouring nerves and blood-vessels.

334. The principal *pivot* joint in the body is that formed by the *odontoid process* of the *axis*, and the *anterior arch* and *transverse process* of the *atlas*. The former is a vertical *peg*, which fits into a ring formed by the two latter, the *ring* of the *atlas* rotating.

335. A Lever is usually defined as a rigid or inflexible rod or bar, movable on or about a certain *fixed* or *relatively fixed* point of *rest*, prop, or support, termed the *fulcrum*. The force by which the lever is moved is termed the *power;* the resistance to be overcome is termed the *weight* or *resistance.*

Levers are divided into *three classes* or orders, according to the position of the points on the bar or lever to which the *power* or the *weight* are applied in relation to the *fulcrum.* The rods in all the three classes of levers may be either *straight* or *curved.*

336. In Levers of the First Kind the *fulcrum* is placed *between* the power and the weight, which are therefore placed on *opposite* sides of the *fulcrum.* The beam of an ordinary balance (pair of scales), a *pump handle,* the arms and blades of a pair of scissors, are so many familiar illustrations of this order of lever. The following are examples of this kind of lever in the human body :—

1. The head rocking backward and forward on the *atlas* (its

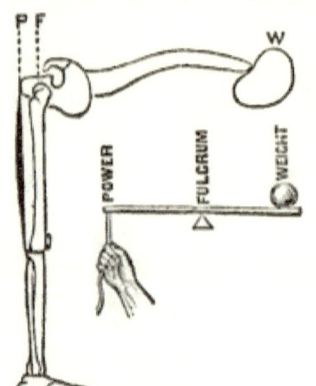

fulcrum), the *trapezium* muscle, attached behind to the occipital bone, being the *power;* the weight of the cranium and the face in front of the *atlas* being the *resistance.* (When a person goes to sleep soundly, or dies in a sitting posture, the head *falls forward,* because the *trapesium* muscle (the *power*) suddenly ceases to act.) 2. The *pelvis,* supported by the heads of the *femoral* (thigh) bones, when raising the trunk from the stooping position, as when bent forwards with the face to the ground (see fig. 66).

Fig. 66. Lever of the First order.   337. In Levers of the Second kind the power and the *weight* act on the *same* side of the *fulcrum,* the *weight* being the nearer. A pair of nut-crackers; a loaded wheelbarrow resting on its wheel (the fulcrum), its handles being raised up; the oar of a boat in the act of rowing.

The following are illustrations of this kind of lever in the human body :—

1. The bones of the foot, when we stand "tip-toe," the toes the *fulcrum;* the *ankle joint* and *body resting* on it the weight; and the *gastrocnemius* muscle, pulling at the *os calcis,* the power (see Fig. 67).

2. The lower jaw in opening (pulling down) the mouth.

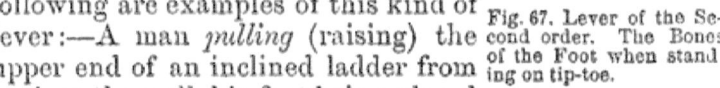

338. In Levers of the Third kind the *power* and the *weight* act on the *same* side of the *fulcrum;* but the *power* is in this case the nearer. The following are examples of this kind of lever:—A man *pulling* (raising) the upper end of an inclined ladder from against the wall, his foot being placed against the *foot of the ladder* as fulcrum (see fig. 68); a pair of common fire-tongs used to hold a lump of coal; the treadle of a lathe.

Fig. 67. Lever of the Second order. The Bones of the Foot when standing on tip-toe.

The following are examples of this order of levers in the human body:—

1. The *radius* of the fore-arm pulled up by the *biceps* muscle.

2. The *bones of the foot,* when the heel rests on the ground, and the toes are pulled upwards by the muscles at the front of the leg.

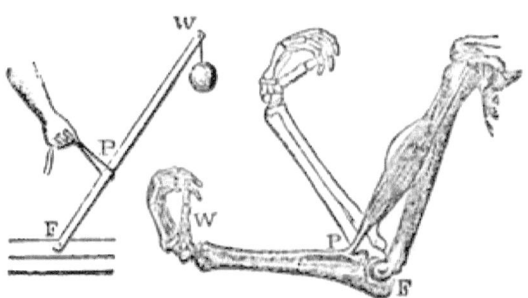

339. The Erect Position of the Body in Standing is

Fig. 68. Lever of the Third Order.

maintained by the complicated *antagonistic* actions of the *voluntary* muscles, which neutralize or *balance* each others *contraction,* as follows:—

(*a*) The muscles of the calf, acting against the foot as the *basis of support* of the body, *contract,* pulling the body *backward,* and thus prevent its falling forward.

(*b*) The *antagonist* muscles of the *front* of the leg and thigh contract, pull the body forwards, and *neutralize* the action of those of the calf.

(*c*) The muscles of the buttocks, spine, and back of the neck *neutralize* or *balance* the forward pulling action of those of the leg and thigh.

(*d*) The muscles of the *front* of the abdomen and throat again

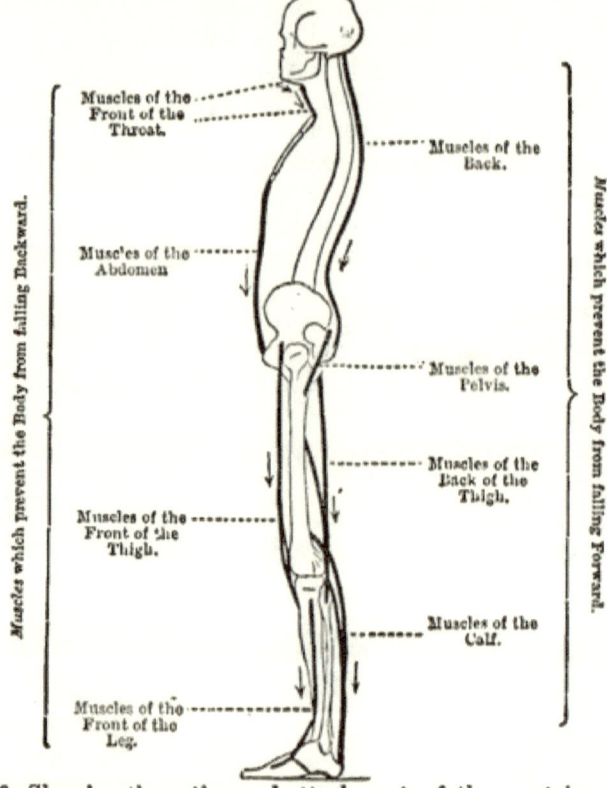

Fig. 69. Showing the action and attachments of the most important muscles by which the Erect Posture of the Body is maintained. The *arrows* show the direction in which the muscles pull, the *feet* acting as the fixed Basis. (After Huxley.)

balance the pulling-backward action of those of the buttocks, spine, and neck.

This mutual balancing of the various *antagonistic* muscles is shown in fig. 69. The arrows show the *direction* in which the muscles tend to pull the body.

In this way the body is kept *erect; its centre of gravity,* however, being *high* up in the trunk, it is, though a process of *almost unconscious* voluntary action, one of considerable delicacy. A blow on the head, a stab or shot, which produces a sufficient *nervous shock* to *suspend* the action of the *will* over the *voluntary* muscles, suddenly arrests their action and causes the body to *fall*.

CHAPTER XVII.

THE ORGANS OF THE VOICE.

340. Voice.—The most delicate and perfect *motor* apparatus in the body is, perhaps, that of the voice : it has been calculated that upwards of 900 movements per minute can be made by the movable organs of speech during reading, speaking, singing, &c. All sound is sensation, produced by the rapid *vibration* of air, or some highly *elastic* medium. *Voice* is *sound* produced by sonorous vibrations, or aerial sound-waves, excited by the rapid vibration of the *true vocal cords*, themselves put into vibration by the rush of air expelled during *expiration* through the *glottis* or narrow chink left between them when they are *tightly stretched*.

341. The student will have observed that if clothes lines or telegraph wires are allowed to swing *loosely* in the wind, no sound is heard ; but that if drawn very tight, they will emit a *musical sound* with every gust of wind. Such is precisely the case with the *vocal cords ;* if they be allowed to hang *loosely* while we *breathe,* no sound will be heard ; but if they are suddenly drawn tight, the mode of breathing being in every other respect unchanged, sound (voice) will be immediately heard. The principal organs of the voice are the *larynx*, or voice-box, and the included *vocal cords.*

342. Speech is ordinarily *voice* carved, chiselled out, or modified into words by the tongue, lips, teeth, palate, cheeks, nose, &c. ; but there may be *speech* without *voice*, as in whispering, in which the *vocal cords* play no part. So, also, there may be voice without *speech*, as when we simply breathe with the *vocal cords* in a state of *tension*.

343. **The Larynx** (from Gr. *larugx*, orifice of the wind-pipe) is the somewhat complex *funnel*-shaped structure at the top of the *trachea*. It is situated immediately in *front* of the upper part of the *œsophagus* and under the tongue. Its *thyroid* cartilage forms the well-known prominence so strongly marked in some men at the upper part of the throat, termed the *pomum Adami*, or Adam's apple ; so named, it is said, because when Eve gave Adam some of the forbidden fruit, a portion of it probably stuck in his throat and produced the swelling or enlargement referred to. (See fig. 70.)

344. **Structure of the Larynx.**—The larynx comprises the following essential parts or structures :—1. A tubular or *funnel*-shaped cartilaginous box or framework. 2. Two *elastic* ligaments, bands, or cushions of *yellow elastic tissue*, situated one on each side of the larynx, separated from each other by an opening in the middle

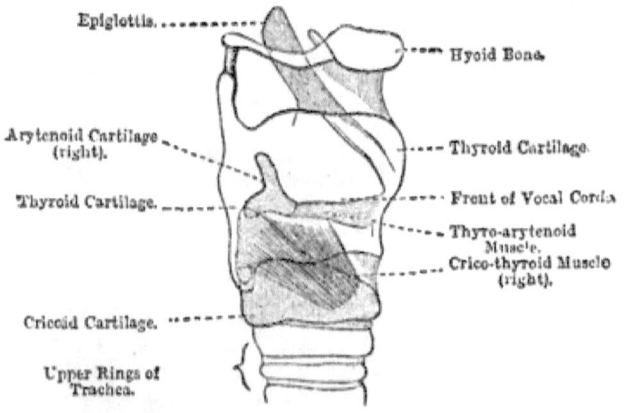

Fig. 70. Plan of Larynx and its Cartilages.
The Thyroid Cartilage is supposed to be transparent, so that the Arytenoid Cartilage, Vocal Cord, Cricoid Cartilage, Thyro-arytenoid Muscle and Epiglottis are to be seen through it.

of the larynx, between the two bands, which are termed the vocal *cords*. 3. Muscles for giving movement to cartilages, and thus *tightening* or *relaxing* the *vocal cords*, so that they may either be put *into* or *out* of action at the command of the *will*, or by which the *sound* they produce may be modified as desired.

It is lined on its interior with *mucous* membrane and is abundantly supplied with nerves, chiefly derived from the *pneumogastric* nerve.

345. Cartilages of the Larynx. — The larynx is built up of four principal cartilages—viz., the thyroid, cricoid, and two arytenoid cartilages.

346. The Epiglottis is the thin elastic, yellowish, leaf-shaped plate of *fibro-cartilage* attached to the thyroid cartilage and the hyoid bone which

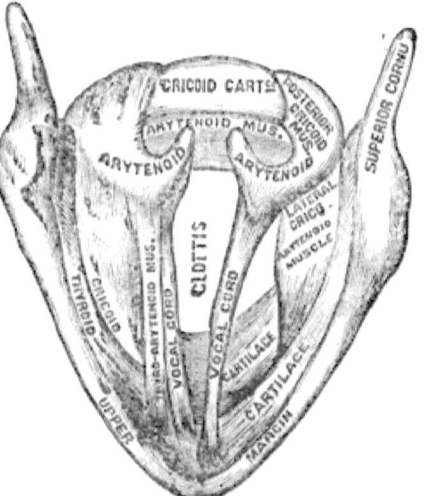

Fig. 71. Bird's eye view of the top of Larynx.

forms a sort of trap-door valve protecting the entrance to the larynx. (See *Deglutition*.)

347. The True Vocal Cords or Ligaments are the two *elastic* bands or cushions of yellow elastic tissue previously described: their surfaces are lined by *mucous* membrane. They are somewhat triangular in their cross section, the bases of the triangles forming the inner edges towards the sides of the larynx, and the apices, the free edges between which the *glottis* (the opening to the windpipe) lies.

Their *front ends* are inserted in the notch in the interior of the front of the *thyroid* cartilage (see fig. 71), their *back* ends being attached to the bases of the movable *arytenoid* cartilages.

348. Essentials to the production of Voice.—From what has been shown, it will be seen that the following conditions are essential to the production of the human voice :—

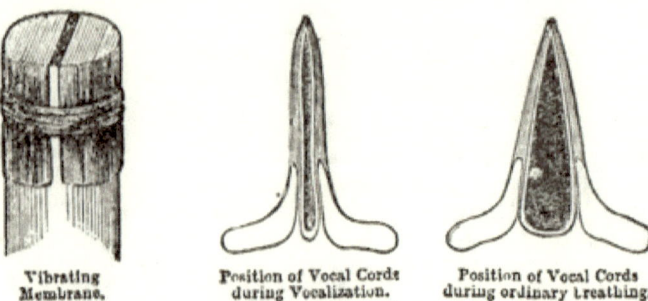

Vibrating
Membrane.   Position of Vocal Cords
during Vocalization.   Position of Vocal Cords
during ordinary breathing.

Fig. 72.

1. The existence of *parallel* elastic ligaments (vocal cords) in the requisite state of *tension.*
2. The passage through the *parallel chink* between these cords of a current of air moving with a sufficient force to put them into the requisite state of *vibration*—that is, to cause them neither to vibrate too *slowly* nor too *rapidly.*

CHAPTER XVIII.

THE SENSES AND THE ORGANS OF THE SENSES.

349. The Organs of the Senses are the instruments by which the mind is brought into *relation* with the *external* world ; or, in other words, the instruments by which the mind is acted upon by natural agencies *external* to the brain.   They consist essentially of *nerve expansions* (spread out and specially prepared to receive the stimulus of the particular agent, mechanical, optical, sonorous, olfactory, or gustatory), which are in general connected with the *brain* by special nerves.   The nature of the sensation depends to a great extent upon the nature of the *covering* of the nerve expansion intervening between the terminal nervous *network* and the external exciting agent.

350. There are six senses, viz. :—the *muscular* sense, the sense of *touch*, of *taste*, of *smell*, of *hearing*, and of *sight*.

In all cases sensation takes place in the *brain*, and not in the nerves or their outer extremities.

351. The Muscular Sense is the sense by which we judge of the *relative weight* of a body, or the *degree of resistance* it offers to effort made to put it into movement.

352. The Sense of Touch, or the sense by which we become acquainted with the existence, shape, and properties of bodies, is common to the whole body, but more especially to the *skin*, some portions of which are very much more highly endowed with this power than others. (See *Skin*, secs. 308–10.)

It is by this sense chiefly we get a notion of *solidity* and *roundness*. In this sense it is the corrective of sight, by which, until corrected by "touch," aided by experience and judgment, all objects appear flat. To persons born blind, and whose sight has been first obtained by the aid of the oculist, all objects at first appear as though they were *flat* and *touching* the eyes.

353. The organ of *touch* consists essentially of an *external* layer of *epithelium*, which, being in contact with the external agent, first receives and modifies its action, which it then transmits to the *internal* layer of the *tactile* organ (consisting of plexuses of nerve fibrils) immediately below it, which transmit the *stimulus* thus originated to the brain, by means of the *cerebral*, or the posterior branches (sensory) of the *spinal* nerves.

The nature of the sensation, to a great extent, depends on the thickness of the medium or covering external to the nerves. If, for instance, the *cuticle* be abraded, and the skin below be touched with a point (however gently) instead of the ordinary sensation of touch, that of *pain* will be produced.

354. **The Tongue** is the chief organ of Taste; but this power is also possessed by the back of the *palate* and the fauces. The tongue consists essentially of a mass of voluntary muscular fibre, covered externally with a layer of *mucous membrane,* in which the sense of taste resides. It is divided by a median line into two lateral symmetrical halves, and has a tip, border or edge, and *dorsum* or back (see fig. 53). Taste is an exceedingly complex sensation.

The *mucous membrane* of the tongue is studded with *papillæ,* of which there are *three* varieties. (See fig. 53.):—

The *larger* papillæ are very vascular, and receive *nerve fibrils* from the *glossopharyngeal* and the *fifth* pair

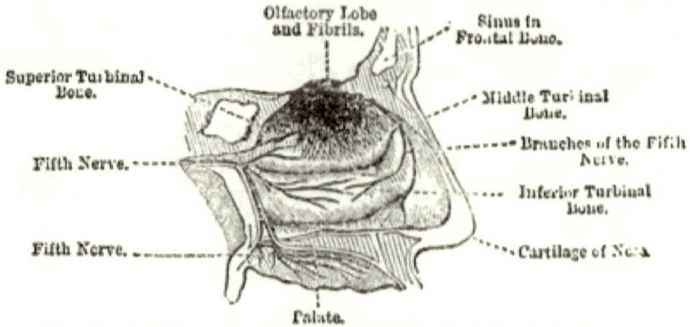

Fig. 73. Vertical Longitudinal Section of the Nasal Cavity. Showing Olfactory Lobe and distribution of the Olfactory Filaments, and the Fifth (Trigeminal) Nerve on the Right wall of the Nose.

of nerves. The former supplies the *back* of the *tongue* and *palate,* which is the chief region of taste; the latter chiefly supplies the *front* of the tongue.

355. **The Sense of Smell** is exercised through the *unciliated mucous* (Schneiderian) membrane which lines the upper parts of the *nasal cavities,* and which receives its supply of nerve-filaments from the *olfactory lobes* and *not* from the fifth pair of nerves.

It is excited by the contact of odoriferous particles, in all probability undergoing the movements involved in the process of *oxidization;* since (1) *odorous* bodies are *oxidizable;* and (2) *no* sensation of *smell* can be excited if *oxygen* be shut off from the nose.

356. **The Nose** is the triangular-shaped organ situated in the middle of the face. Its *roof* is formed by the *cribriform plate* of the *ethmoid* bone of the skull, through the *sieve-like* apertures of which the *olfactory filaments* (false nerves) pass. It is bounded in front and laterally by the *nasal bones* and *cartilages;* its floor is formed by the hard and soft palate. It is divided into *two cavities* by the *nasal septum* (which consists partly of bone, the *vomer*, and partly of *cartilage*). These cavities open out into the air in front of the nose by means of the two *nostrils*, and into the *pharynx* behind, by the two *pos-*

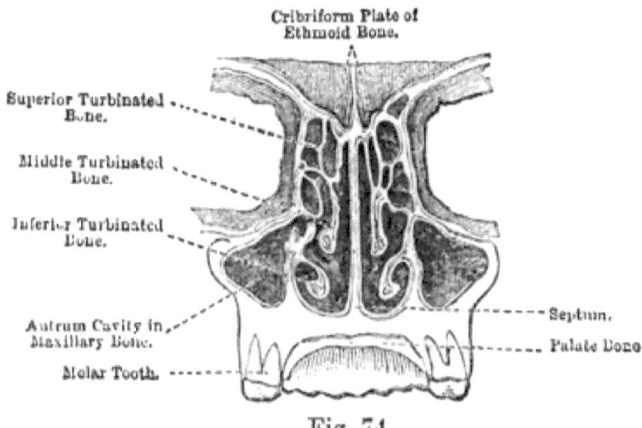

Fig. 74.

terior nares (nostrils) situated immediately over the sides of the *velum* or soft palate.

357. **Sound.**—The *external* cause of sound is mere *mechanical* movement. Sound is almost invariably produced by *air* in a state of *sonorous vibration*, that is, air oscillating *backwards* and *forwards* with great rapidity. If the *wave movement* be either too *quick* or too *slow*, it will not produce sound.

358. **The Sensation of Hearing** is excited in the brain by means of a *molecular movement*, set up in the *nerve fibrils* of the *internal ear* or labrynth, by the rapid vibration of some external *elastic* body, and transmitted to the brain by the *auditory nerve*. The essential parts of

the organs of hearing are the *membranous labyrinth*, and the *scala media* of the *cochlea*.

359. **The Organs of Hearing** (the ears)—each consist of three parts, viz. :—

(1.) **The external ear**, comprising the *pinna* or auricle, the gristly appendage attached to the side of the head,

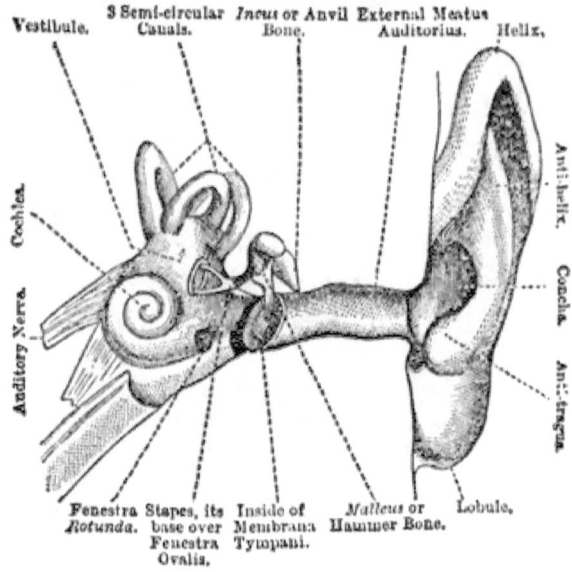

Fig. 75. Diagram of Ear.

(which both serves as a natural ornament and to collect the vibrations of the air,) and the *auditory canal* (meatus) or passage by which the vibrating air is conducted to the membrane of the *tympanum*. The meatus is lined by mucous membrane, studded with the *ceruminous* or wax glands.

(2.) **The tympanum**, or *middle ear*, (which consists of an irregular cavity in the *petrous* part of the *temporal bone*), bounded on its outer side by the *membrana tympani*, and on its *inner* side by the outer wall of the *bony labyrinth*. It is traversed by a chain of movable bones, consisting of the *malleus* or hammer bone, the *incus* or anvil bone, the *stapes* or stirrup bone, by which the vibrations are conveyed from the external air, through the middle ear, to the membrane in the *fenestra ovalis* in the side of the labyrinth. The tympanum opens into the *pharynx* by the *Eustachian tube;* by this arrangement the *air* enclosed in the tympanum is kept at the same *tension* or pressure as that of the external atmosphere.

(3.) **The labyrinth**, or internal ear, consisting of the *vestibule*, the *three semi-circular canals*, and the *cochlea*, and their *membranous*, nerve, fluid, and other contents.

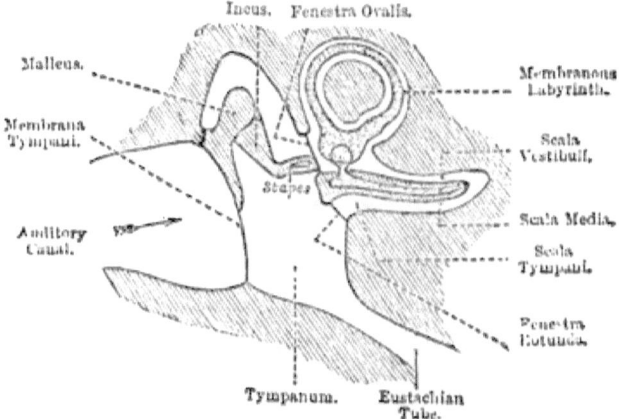

Fig. 76. Plan of Ear.

The Scalæ of the Cochlea are supposed to be unrolled. The shaded portion represents Bone.

360. **The Vestibule** is the middle or central chamber of the *internal ear* or *labyrinth* which opens into the *cochlea* and the *semicircular canals*. It is situated at the inner side of the tympanum, with which it communicates by means of two *membrane-stopped* openings, viz. :—the *fenestra ovalis* and the *fenestra rotunda*.

The *vestibule* contains a *larger* membranous *sac* termed the *utricle*, and a *smaller* one termed the *saccule*, or frequently the *sacculus hemisphericus*.

361. **The Semicircular Canals** are three long arched tubes about $\frac{1}{20}$ of an inch in diameter. These three hollow arches, which form the greater parts of circles

14 E,

of two nearly *vertical* canals (an *anterior* and a *posterior*) and a *horizontal* canal.

362. **The Membranous Labyrinth** consists of a closed membranous sac of the same form as, and a little smaller than, the *vestibule* and the three *semicircular* canals by which it is enclosed. It occupies the middle of these bony structures, and is separated from them by a clear *liquid* termed the *perilymph*, its interior being filled with a similar liquid, termed the *endolymph*. The *vestibular* portion of the *membranous labyrinth* consists of two sacs, —a larger, termed the *utricle*, and a smaller, termed the *saccule*. Each canal has a larger or dilated end, termed its *ampulla*, the nerves of which are covered with delicate stiff filaments. The fibres of the *auditory nerve* are distributed over the inner walls of the *ampullæ*, the *saccule*, and the *utricle*.

363. **Otoconia.**—In order to increase the effect of the *vibratile concussion* on the auditory *nerve filaments*, little masses of minute crystalline grains of stone (carbonate of lime), termed *otoliths*, or *otoconia*, are supplied to the walls of the *saccule* and *utricle*, opposite the point where the nerves are distributed.

364. **The Cochlea** (Lat. a snail's shell), a conical shell-like structure, forms the front portion of the *labyrinth*. It possesses a central *axis*, termed the *modiolus*, round which a partition (partly of bone, partly of membrane), termed the *lamina spiralis*, winds spirally, $2\frac{1}{2}$ times, dividing the spiral canal of the cochlea into two *scalæ* or passages, termed respectively the *scala vestibuli* and *scala tympani*. Between these two passages is a third, termed the *scala media*, which is contained between the two walls of the membranous portion of the *lamina spiralis*. The latter is really a membranous *bag*, twisted *spirally* round the edge of the bony portion of the *lamina spiralis*, and its cavity forms the *scala media*, as described. One of its walls, more elastic than the other, is covered over with minute rod-like bodies, termed the *fibres of Corti*, which, looking like so many keys on a keyboard, serve more readily to take up the vibrations communi-

cated to the *endolymph*. The *interior* of the walls of
the *scala media* are covered with fibres of the *auditory*
nerve. One end of the *scala media* is closed, the other
opens into the *sacculus hemisphericus.*

365. The Modus Operandi of Hearing.—The air is put
into rapid *sonorous vibration*, the aerial waves enter the
external auditory canal, impinge upon the *membrana
tympani*, and put it into the same rate of vibration; the
malleus, pressing against its interior side, is put into
vibration by the *membrana tympani;* the *malleus* puts
the *incus* into vibration; the *incus* attached to the
stapes, puts it into vibration; the *stapes* attached to
the *membrane* filling up the *fenestra ovalis*, an oval
aperture in the vestibule, puts it into vibration; this
membrane puts the *endolymph* (in the interior of the
semicircular canals, ampullae, saccule, utricle, and *scala
media*) into vibration; the *endolymph*, dashing against
the auditory nerve *fibrils, otoliths, otoconia,* and *fibres
of Corti*, puts them into vibration, the *fibres of Corti*, like
the keys on a pianoforte, only taking up the vibrations
corresponding to their length and special note. The
vibrations thus set up *synchronously* with the external
vibrating air, acting as an *excitant* on the *auditory nerve,*
cause it to transmit to the *brain* a *nerve-movement* or
stimulus which wakens up in it the *sensation of sound.*
A little membrane-stopped hole, the *fenestra rotunda* in
the vestibule, facilitates the *movements* of the *endolymph,*
by giving it more play.

366. Light is the *external* agent or *cause* of *normal* or
objective vision. Of its real nature we *know* absolutely
nothing, all our knowledge of it being purely *hypo-
thetical.* Yet in the whole range of human knowledge
we possess no explanation of any of the phenomena of
nature more complete and satisfactory—if even so complete and satisfactory as—that afforded by the *undulatory
hypothesis of light*, of the various complex and beautiful
phenomena of light and colour.

367. Undulatory Hypothesis of Light and Colour.—This
hypothesis assumes that all *planetary space*, also all

interstitial space (the pores or space between the particles of matter), is filled with a highly attenuated, *imponderable*, invisible, *elastic* fluid, termed *luminiferous ether*. It also assumes that this *ether* is capable of being put into an up-and-down *wave-movement*, which in *direction* and general character resembles that of the sea, but which, in the *minuteness* of the waves, and the rapidity of the propagation of their movement, is utterly inconceivable to the human mind. It is supposed that these waves come "rolling in" through the *openings* and *transparent humours* of the eye, *pitching* themselves against the *retina* at the back of the eye, like sea-waves pitching against the *rocks* or against a *sea-wall*. It is further supposed that the nerve fibrils of the retina are shaken by the *wave concussions* they thus receive into a series of vibrations, that these vibrations act as a *stimulus*, which, transmitted to the *brain* by the *optic nerve*, produces the *sensation* of sight.

It is further supposed, that if the nerve fibres of the *retina* receive 390 *millions* of millions per second of wave *concussions*, they will themselves be put into the same rapid *rate of vibration*, and stimulate the brain, through the medium of the *optic nerve*, so as to cause a *sensation* of just visible *redness;* that if they are put into vibration by *shorter* light-producing waves, travelling at the rate of 754 millions of millions of *waves* per second, they will produce a just visible sensation of *blue* or *violet* colour.

If the *nerves* of the retina be made to *vibrate* at intermediate rates to the above, one or other of the colours of the spectrum or rainbow will be produced; but if the rate of vibration be either *very much higher* or *lower* than those given, no sensation of *light* or *colour* will be experienced.

368. Colour, like sound, is thus a sensation consequent on *brain-change* produced by the *transmission* of a *molecular movement* or change in the *substance* of the nerve fibrils to the brain, the movement being *originated* by an external *vibrating agent*. In the case of sound, the external agent is the *air;* in the case of light or colour, it is the *luminiferous ether.*

369. **Colour Blindness,** or **Daltonism,** consists in the inability of certain eyes to distinguish particular colours. Singularly the most common defect of this is the inability to distinguish *red* from *black, green,* &c.

It is not yet known whether this weakness arises from a defect in the *brain,* the *retina,* or the *humours* of the eye. The employment of *colour-blind* persons as railway guards might lead to most serious accidents.

370. **The Eye** is essentially an *optical* instrument, constructed for receiving, *bending* (refracting), and throw-

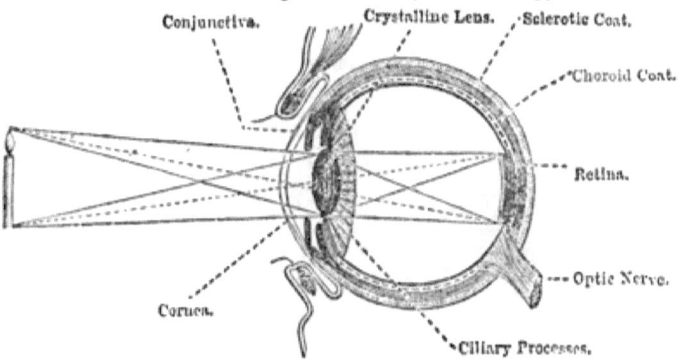

Fig. 77. Showing the formation of *inverted* optical images on the Retina at the back of the Eye.

ing the rays of light on to a screen (the retina) at its back, so that it shall receive a very *minute* and *inverted,* but clear and definite, *picture* or *image* of the surrounding objects. In fact, in no case do we see the external objects themselves, but *pictures of them formed by the light* sent from them, and *focussed* on the back of the eye (the retina), as just described. The eye is, in fact, a sort of water *camera obscura:* it is moved by *six* muscles attached to its external coat (the *sclerotic*). The eyes are lodged, for protection, in packings of fat in the *orbits* of the cranium.

371. Structure of the Eye.—The eye is a nearly round ball, about 1 inch in diameter, which encloses *three* lenses or *humours* and *two* muscles, and which consist of three *coats* or layers. It also contains nerves and blood-vessels. It is attached to the *optic nerve* behind, as an apple to its stalk.

Coats of the Eye.	Refracting Humours.	Muscles.
1. (Outer) Sclerotic and cornea,	Aqueous,	The Iris.
2. (Middle) Iris, ciliary, and choroid,	Crystalline (lens),	Ciliary muscle.
3. (Inner) Retina,	Vitreous.	

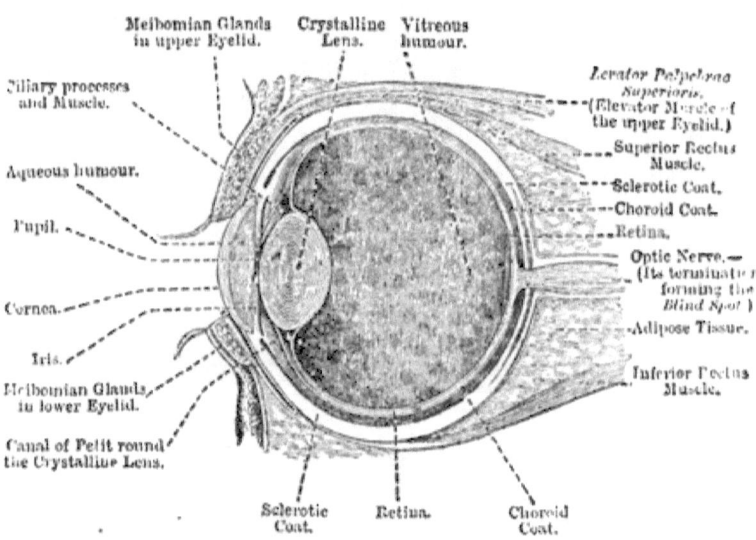

Fig. 78. Vertical Section of Eye-Ball.

372. The Sclerotic Coat (from Gr. *skleros,* hard) forms the external wall of the eye-ball, the "white of the eye." It is an opaque, tough, fibrous membrane, which cuts like leather, and consists chiefly of white fibrous tissue. It contains two apertures—the circular opening in front, in which the *cornea* is inserted, and the *posterior* opening through which the *optic nerve* enters. (See fig. 78.)

373. The Cornea (from Lat. *cornu,* horn) is the circular watch-glass shaped, transparent, fibrous body inserted in the aperture in the *sclerotic coat*, at the *front* of the eye, that admits the *light* by which vision is excited. (See fig. 78.) It also aids in bending or focussing the light which enters the eye.

374. **The Choroid Coat** (from Gr. *chorion*, the outer skin of the egg) is the delicate coat of *blood-vessels* and *black* pigment cells which form the *middle* coat of the eye, and causes the black appearance of the pupil. When the pigment is *wanting*, as in the case of Albinoes, the blood-vessels, showing through the aperture of the pupil, give it a *red* or *pinkish* appearance. Towards the front of the eye it collects into about *sixty folds*, which are termed the *ciliary processes*, to which the *iris* is attached by a narrow fibrous ring, termed the *ciliary ligament*. (See fig. 78).

375. **The Iris** (from Lat. *rainbow*), so called from the diversity of its colour, is the *circular*, flattened, perforated *curtain* of *unstriped muscle* nerve, connective tissue, pigment cell, and blood-vessels, which, placed *behind* the *cornea*, *regulates* by the contraction and expansion of its central aperture (the *pupil*) the quantity of light admitted to the eye. It divides the space between the *crystalline lens* and cornea, which contains the *aqueous humour*, into an *anterior* and a *posterior* chamber.

EXPERIMENT.—Place yourself before a looking-glass in a *dark* room with a *lighted* candle in your hand, hold the light as far away to the side as you can, while you look at the image of the *pupil* of your eye. It will appear very *large* and *dark*. Now bring the candle gradually nearer and nearer until you bring it close before the eye—the pupil becomes *smaller* and *smaller* because of the contraction of the circular muscular fibre of the iris.

376. **The Aqueous Humour** is the clear, limpid, watery fluid which fills the space in *front* of the crystalline lens, and bathes both sides of the *iris*.

377. **The Crystalline Lens or Humour** is the biconvex lens-shaped, transparent, jelly-like body, placed almost immediately *behind* the *iris*, by which the light entering

the eye is *focused* and made to form *inverted pictures* or
images on the *retina* at the back of the eye. It is
about ⅓ of an inch in diameter and ⅛ of an inch thick.
Its general form and properties may be well studied in
the eye of a sheep. It is retained in its position by the
suspensory ligament, and is encircled by a triangular
cavity, termed the *canal* of *Petit*, which probably gives
space for its *adjustment*.

378. **The Vitreous Humour** is the large, spherical, transparent, *glassy-looking* lens or humour which fills up the
greater part of the interior of the eyeball. It consists
of a jelly-like albuminoid fluid, inclosed in a delicate
capsule, termed the *hyaloid* membrane. (See figs. 78.)

379. **The Retina** (from Lat. *rete*, network) is the delicate coat or membrane which may be seen lining the
interior of the back of the eye, when the eyeball is carefully cut into a *front* and a *back* half. It consists partly
of an *expansion* of the *optic nerve*, and partly of *other*
structures, which probably assist in enabling the *light*
to produce the requisite impression on the *nervous
fibrils* of the optic nerve.

380. **The Blind Spot**, optic pore or *punctum caecum*,
is the *insensible* (to light) portion of the *retina*, situated
at the back of the eye at the *entrance* of the *optic nerve*
and *central artery*. The *optic nerve* enters the eye a
little inside (towards the nose) of the *optic axis* or
line which passes *perpendicularly* through the *centre* of
the *crystalline lens.*

Images of objects falling on the *blind spots* of the eyes
are quite *invisible.*

EXPERIMENT (1.) Hold the book so that the letters A and B
shall be 9 or 10 inches or thereabout from the eyes.
(2.) Shut the *right* eye and look *continuously* and *steadily* at the
letter B on the *right*. The letter A will also be seen.
(3.) Move the book slowly towards the eye, taking great care
not to alter the *direction* in which it looks. At a certain
point the letter A will disappear, its image is now on the
blind spot.
(4.) *Continue* to move it *towards* the eye, and its *image* will be
removed from the *blind spot* to a *sensitive* part of the
retina, and *both* letters will again come into sight.

A   **B**

381. **The Bright Spot of Sömmering,** *macula lutea,* or yellow spot, is a round, yellowish, elevated spot, about $\frac{1}{24}$ of an inch in diameter, situated in the *centre* of the back of the eye in the *axis of vision,* and about $\frac{1}{10}$ of an inch *outside* of the *blind spot.* Its *summit* contains a little pit or *depression* termed the *fovea centralis.* It is the seat of most *acute vision,* yet it has *no nerve fibres* from the optic nerve, but it is full of close-set *cones,* and contains *nerve corpuscles.*

382. **The Duration of the Impression of Light** on the retina is about $\frac{1}{8}$ of a second. If, therefore, a lighted stick be rapidly moved round in a circle so that it shall return to the point from which it started in less than $\frac{1}{8}$ of a second, it will be seen as though it were a *luminous circle.* The appearance of the firework termed the "Catherine wheel," and of the pictures in a *zoetrope,* are due to this cause.

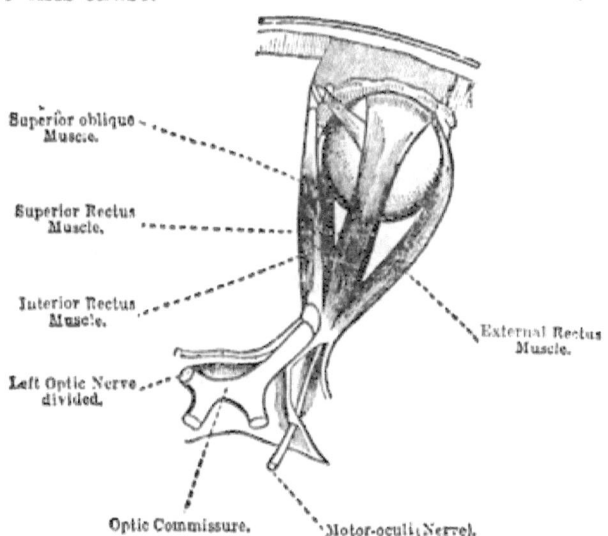

Fig. 79. The Muscles of the Eye-Ball.

383. **The Muscles of the Eye-Ball** by which the eyes are rolled about and *axis of vision* is moved in any given direction, consist of the four *recti* (straight) muscles by

which the eye is rolled inwards, outwards, upwards, and downwards, the *superior oblique* pulley or *trochlearis* muscle, and the *inferior oblique* muscle by which the eye is *rolled on its axis* at the same time that it is pulled *inward* and *forward*. The *two oblique* muscles are attached a little behind the centre on the *outer* side of the eye-ball, and thus give it their peculiar movement. The tendon of the *trochlearis* muscle passes through a tendinous *pulley-like* loop, and bends downwards so as to act on the eye-ball like a cord from a pulley. The following are the names and functions of the muscles:—

1. Superior rectus (*attolens*) muscle, pulls the eye-ball upwards.
3. Inferior rectus,   ,,   ,,   ,,   downwards.
3. Internal rectus (*adductor*), ,,   ,,   ,,   inwards.
4. External rectus (*abductor*), ,,   ,,   ,,   outwards.
5. Superior oblique (*trochlearis*), rotates the eye out-
    ward and downward.
6. Inferior oblique, rotates the eye outward and upward.

384. The Chief Appendages of the Eye are the eyebrows, eyelids, conjunctiva, and the lachrymal apparatus.

385. The Eyebrows are the arched integumentary prominences which project over the upper part of the front of the orbits. Among their other offices they shade and protect the eyes, and with the aid of the short thick hairs with which they are studded, prevent the perspiration from running into them.

386. The Eyelids consist of thin plates of movable cartilage surrounded by folds of skin. Each of their *free* edges is fringed with a row of hairs (the eyelashes), and contains a row of from 20 to 30 minute glands termed the *Meibomian glands* (see fig. 78), which consist of modified *sebaceous glands* embedded in grooves in the cartilage. Each of these glands consists of a single closed straight tube of *basement membrane*, into the sides of which a number of minute follicles open; the interior of the gland is lined with scaly epithelium. The *upper* eyelid is *raised* by the contraction of a special muscle termed the *levator palpebrarum superioris*. The eyelids are *closed* by the contraction of a sort of *sphincter* muscle termed the *orbicularis palpebrarum* muscle.

387. **The Lachrymal Glands** consist of two small *racemose* glands (each about the size of an almond), lodged in depressions at the upper and outer angles of the orbits (see fig. 80). They secrete the *lachrymal* fluid which *moistens* and *lubricates* the front of the eye,

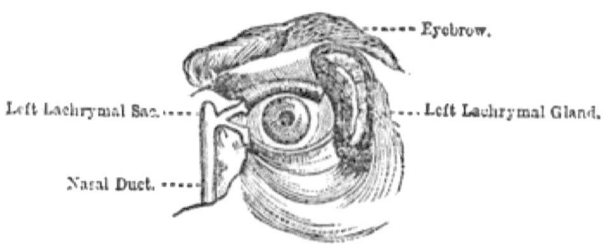

Fig. 80. The Lachrymal Glands.

and which passes off from the inner angles of the eye by the *lachrymal* and *nasal* ducts into the nose. When secreted in very large quantities, as during certain kinds of mental excitement or in consequence of the action of *irritants*, a part of it escapes as *tears* down the cheeks.

CHAPTER XIX.

THE NERVOUS SYSTEM — INNERVATION.

388. **The Nervous System** consists of the cerebro-spinal axis which comprises the *brain, medulla oblongata, spinal cord,* and the *cerebral* and *spinal* nerves, and of the *sympathetic, ganglionic,* or *organic nerve* system.

The Brain and Spinal cord are enclosed within three coverings. The *dura mater,* an outer tough, fibrous membrane, which also lines the skull—a middle *serous* membrane, the *arachnoid* membrane—and an inner *vascular* membrane, the *pia* mater, which adheres to the brain dipping into its fissures.

389. **Innervation.**—The various functions of the ner-

vous system constitutes that of *innervation*, and consists in the generation and transmission of *motor impulses* (sec. 149), of sensation, and of thought, volition, and emotion. For every act of innervation—that is, for every idea thought, every emotion excited, every sensation felt, *brain tissue* is *burnt* or *oxidized*.

. **390. Sensation** is the process by which we become *conscious* through the *brain* of *impressions* received and transmitted to it by the *afferent* or *sensory* nerves (sec. 398). When sensation is excited *normally*—that is, by *external* agency—it is termed *objective* sensation; but when it arises without any *external* cause, that is, is produced by the unprompted or rather *intrinsic* action of the brain or nervous system itself—it is termed *subjective* sensation, as in the case of the "ringing in the ears" sensation with which most are more or less familiar. Sensation requires—

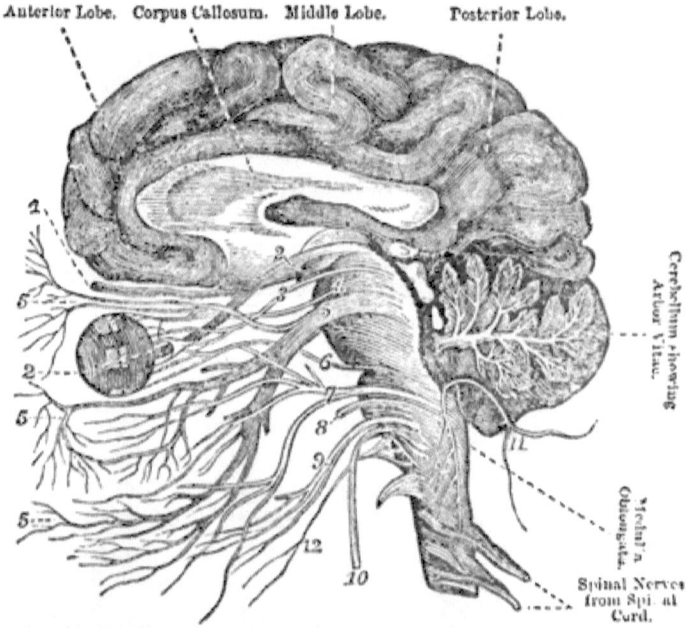

Fig. 81. Side View of Human Brain, showing Cerebral Lobes and Cranial Nerves (of Right Hemisphere), Cerebellum, Medulla Oblongata, and Corpus Callosum.

The observer is supposed to be looking at the right side of the great Longitudinal Fissure, and the *cut portion* of the *Corpus Callosum*.

(1.) A suitable medium for *receiving* the external impression or stimulus—as the eye to receive light.

(2.) A means of *transmitting* the impression to the brain—as the optic nerve.

(3.) Brain organization to develop *consciousness* of impression.

391. The Brain or Encephalon.—The principal parts of the brain are the *cerebrum* or brain proper, the *cerebellum* or lesser brain, the *pons Varolii* and the *medulla oblongata*. It also contains a series of *ganglia* at its base—viz., the *corpora striata*, *optic thalami* separated from each other by the third ventricle, *corpora quadrigemina*, the *pineal gland*, and the *pituitary body* the functions of which are not all understood. It also contains *fissures* or *cavities* within or at its base termed *ventricles*.

The average weight of a man's brain is 54 ounces, and that of a woman's 45 ounces. The maximum weight known is 64 ounces.

392. The Cerebrum or principal mass of the brain is divided by the great *longitudinal fissure* into two hemispheres—each hemisphere is again divided into three lobes—anterior, middle, and posterior—the anterior lobe lies in front of the fissure of Sylvius, the posterior lies over the *cerebellum*, and the middle lobe between the two.

393. Function of the Cerebrum.—That the *cerebrum* is the principal seat of the intellect, volition, and of the emotions, is shown by the following facts:—When the human *cerebrum* is below a given size, its possessor is always an *idiot*. *Disease* or *injury* produces *idiocy* or *insanity*. The *size* of the *cerebrum*, its quality being equal also, bears some proportion to the *mental power* of the animal.

That these powers are derived from the *cortical* or *external vesicular* structure is additionally shown by the fact that in serious slow-growing disease affecting the whole of the brain, that if the disease *first* attack the white internal *medullary* portion, that the power of *muscular control* and movement is *first* lost, the *intelligence* being affected *last,* whereas if the *cortical* part of the brain becomes first diseased the *mind* of the patient is first affected, he either becoming *maniacal* or *demented.* If the *cerebrum* be removed from a pigeon or other animal that can stand the *nervous* shock incurred in its removal, it will live and move but will show no signs of *consciousness* or *intelligence.*

Poisons as alcohol, opium, &c., which act upon the *cerebrum,* also produce temporary insanity or *loss* of *intelligence,* or of *consciousness.*

394. The Cerebellum or lesser brain, situated at the base of the back of the skull, is separated from the cerebrum by the *tentorium,* a process of the *dura mater* which lines the inside of the skull, forming a *floor* for the cerebrum, and a *roof* for the cerebellum. It consists of alternate *laminæ* of white and gray nerve matter which, when cut *perpendicularly,* presents a peculiar *arborescent* appearance, termed the *arbor vitæ* of the *cerebellum.* Its weight is about $\frac{1}{10}$ of that of the whole brain. (See figs. 81 and 82.)

Its function is not fully known: it however in some way or other regulates and co-ordinates muscular movement. If removed from the head of a pigeon, the pigeon will continue to move backward or round and round, having apparently lost all power of regulating its movements.

395. The Pons Varolii, or bridge of Varolius, is the *commissure* which connects the cerebrum, cerebellum, and medulla oblongata together. It consists mainly of white nerve fibre.

396. The Cranial or Cerebral Nerves are the twelve pairs of nerves which are given off from the *brain* or the *medulla oblongata,* and which pass out of nine foramina (apertures) at the base of the *cranium* (skull). The cranial nerves are *numbered* from *before backwards,* according to the order in which they pass out of the skull. The special names, numbers, functions, and distribution of these nerves are in the following table:—

TABLE OF THE FUNCTION AND DISTRIBUTION OF THE CEREBRAL NERVES.

Pair.	Special Name.	Distribution.	Function.
1st.	Olfactory nerves.	To the *upper* part of the *mucous* membrane of the nose.	Sensory (smell).
2nd.	Optic nerves.	To the inside of the eye-balls.	Sensory (vision).
3rd.	Motores oculi.	To the superior, inferior, and internal rectus muscles of the eyes, and to the elevator muscle of the upper eyelid, and to the *iris*.	Motor.
4th.	Trochlear nerves.	Superior oblique trochlearis muscle.	Motor.
5th.	Trigeminal or trifacial nerves.	From the fourth ventricle, by three divisions, to the eye-ball, orbit, lachrymal gland, skin of the face, muscles of the jaws, and front of the tongue (taste).	Mixed (motor and sensory).
6th.	Abducens nerves.	External rectus muscle of the eye.	Motor.
7th.	Facial nerves, sometimes described as the *portio dura* of the seventh pair.	To nearly all the muscles of the face.	Motor.
8th.	Auditory nerves, sometimes described as the *portio mollis* of the seventh pair.	To the various parts of the labyrinth or inner ear.	Sensory (hearing).
9th.	Glossopharyngeal.	To the tongue and soft palate (taste), and to the pharyngeal muscles.	Mixed (motor and sensory) taste and common sensation.
10th.	Pneumogastric nerves (par vagum).	To the mucous membrane and the muscles of the pharynx, the larynx, and the trachea, and to the lungs, the liver, the stomach, and the heart.	Mixed (sensory and motor).
11th.	Spinal accessory nerves.	From the spinal marrow to the muscles of the neck and back.	Motor.
12th.	Hypo-glossal or *lingual* nerves.	To the muscles of the tongue.	Motor.

The *seventh* and *eighth* pairs of nerves leave the cranium by the same apertures; they have therefore, by some writers, been counted as but one pair, viz., the *seventh*. For a similar reason, the *eleventh* and *twelfth* pairs are also sometimes counted as one pair.

397. The Medulla Oblongata is the *cranial* portion of the *spinal* cord. It is the *nervous* centre of the *respiratory* movements; its injury will therefore cause death by suffocation; irritation of the medulla may also cause stoppage of the action of the heart. It is the seat of the origin of all the true cranial nerves.

398. The Spinal Cord is that portion of the *cerebro-spinal axis* which is contained within the spinal column. It commences at the termination of the *medulla oblongata* and extends from the *foramen magnum* (the large aperture in the occipital bone at the base of the skull) to the *first lumbar vertebra*, where it termi-nates in the *cauda equina*. It is about 16 inches long, and weighs, with its nerves and invest-ing membranes, about 1½ ounces. (See fig. 82.)

399. Functions of the Spinal Cord. —1. The spinal cord transmits the commands of the will directly from the brain to the voluntary muscles by the *motor* nerve

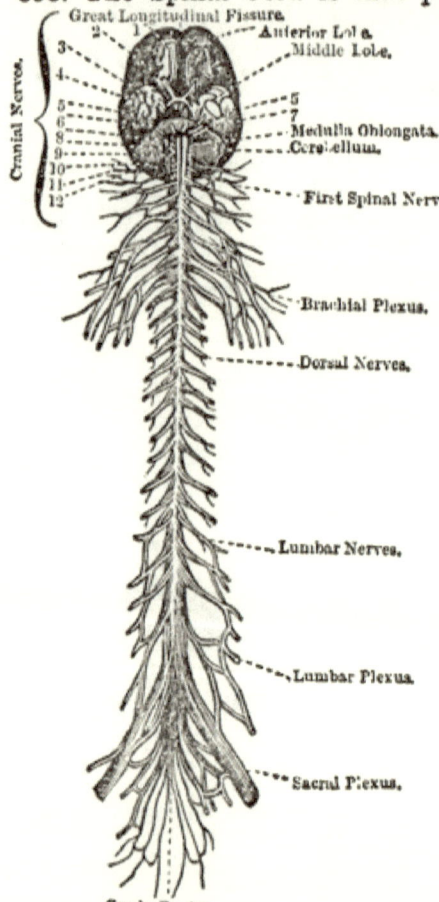

Great Longitudinal Fissure.
Anterior Lobe.
Middle Lobe.
Cranial Nerves.
Medulla Oblongata.
Cerebellum.
First Spinal Nerve.
Brachial Plexus.
Dorsal Nerves.
Lumbar Nerves.
Lumbar Plexus.
Sacral Plexus.
Cauda Equina.

Fig. 82. Showing Human Brain (lower surface), Spinal Column, and Cranial and Spinal Nerves.

fibres it gives off in the spinal nerves. 2. It transmits sen-

sory impressions direct to the brain, where they excite con·sciousness, or *sensation*.   3. It receives sensory impressions by the sensory nerves which it does *not* transmit to the brain, but which, acting as a stimulus, cause *it* to send back the *motor impulses* to the muscles which cause them to contract altogether independently of our consciousness, or of the brain.   This constitutes *reflex* action. It is in this way that the various movements of digestion are carried on.

If a frog be *decapitated*, and its feet or legs be irritated by the point of a needle or a drop of acid, it will kick violently, and will even in some cases, where one leg only is irritated, bring or try to bring its second leg to aid the first in its attempt to get rid of the cause of irritation. In this case both the *brain* and *medulla* being *removed*, there can be no power of *thought* or *consciousness* remaining in the frog. These movements must therefore result from the *reflex* action of the spinal cord.

The *spinal cord* (including the *medulla oblongata*) is thus an *independent* centre of *nervous (reflex) action*, in addition to being the medium by which the brain is brought into nervous connection with the rest of the body. This action is, as previously stated, due to its gray *vesicular* nerve-substance.

400. **Reflex, or Excito-Motor Action.**—If the spine be broken or injured, all parts of the body below the injury become *paralysed* (lose their power of sensation and voluntary movement). If a hot iron be applied to the feet in such a case, the legs will kick out violently, though the patient is quite unconscious of a sense of heat or pain, even tickling the feet will produce this effect.

The sensory (afferent) nerve fibres conduct the stimulus (the irritation) to the spinal cord, which immediately, as it were, *reflects* it back by the *efferent* nerves, in the shape of *motor* impulses, to the legs, which therefore kick unconsciously. *Coughing, sneezing, winking* when an object suddenly approaches the eye, *infantile convulsions*, tetanus (lock-jaw), the *peristaltic* movements of the stomach and intestines, are all so many cases of *reflex* action.

When, from long practice, certain movements at first

14 E,   M

requiring great attention, as those required in playing the violin or the pianoforte, or in painting a portrait, can be executed so easily that the musician or artist can talk and think freely on other subjects while at work, a sort of *artificial reflex action* has been acquired by education. Such operations may also be regarded as instances of *unconscious cerebration.*

· **401. The Spinal Nerves** consist of the thirty-one *pairs* of nerves which pass off from the *sides.* of the entire length of the *spinal cord* leaving the vertebral canal by the *intervertebral foramina* (see sec. 67) on either side of the vertebral column.

Each spinal nerve arises by *two roots*, an *anterior* root, consisting of *motor* nerves, and a *posterior* root, consisting of *sensory* nerve fibres. The *two roots* on each side unite as they leave the spinal cord to form *single trunks*, which shortly subdivide, giving off smaller branches, which ramify through the system. The posterior, afferent, or *sensory* roots have ganglia. (See fig. 82.)

402. Injury or Irritation of the Spinal Nerves.—If a trunk nerve be cut or injured just as it leaves the vertebral column, entire *paralysis* is produced, the power of *sensation* and *motion* being entirely lost to the parts to which the nerve is distributed. If the *motor* nerve only is cut *partial paralysis*, consisting in the loss of all power of *motion* is produced, sensation being retained. If a *sensory* root only is cut, the *paralysis* caused by the injury involves only the loss of the power of *sensation* of the parts to which the nerves are distributed.

If the *end* of a *cut* nerve trunk most *remote* from the spinal column be pinched, or irritated by an electric current from a galvanic battery, it will cause *muscular contraction* or *convulsion* of the parts in which the nerves from that trunk ultimately terminate; but if the *nearer* end to the spinal cord be pinched or irritated it will cause great pain.

If the spinal cord were to be divided *perpendicularly* down its chief fissure by a knife, sensation would be destroyed all over the trunk and limbs.

403. The Sympathetic Nerve System comprises—
1. The *pre-vertebral* double chain of ganglia. 2. The *isolated* ganglia of the viscera, including the *cardiac* (see sec. 200), *hypogastric* and *solar* plexuses. 3. The *ganglia* on the *posterior* roots of the *spinal* nerves.

The *pre-vertebral* ganglia, which form the chief part of this system, consist of two parallel rows or chains of about thirty *ganglia*, situated on each side of the front of the spine. These double rows of ganglia unite together in a *ganglion*, termed the *ganglion impar*, opposite the *os sacrum*. These ganglia are connected with each other, also with the *spinal* nerves, and with the *isolated* ganglia, by means of *gelatinous* and *white nerve-fibre*. Many of these *nerve-fibres* originate in the sympathetic system, others, doubtless, in the spinal cord.

The *great solar* or *epigastric* plexus is situated in the abdomen behind the stomach and immediately in front of the aorta and about the coeliac axis.

The *hypogastric* or *pelvic* plexus is situated in the lower part of the abdomen, chiefly in front of the os sacrum and about the bladder and rectum. It supplies the viscera of the pelvic cavity.

404. The *sympathetic nerves* largely influence the *unstriped* muscular fibres in the walls of the *intestines* and the blood-vessels, and thus *regulate* nutrition. Their *ganglia* are also probably sources of *reflex* action to these organs. Their *motor* nerves, as in the case of the heart (see sec. 200), are, however, under the control or influence of the *pneumogastric* or other *cerebral* or *spinal* nerves. The sympathetic nerve system, most probably to a great extent, though not exclusively, presides over, influences, and co-ordinates the various processes of involuntary motion, of secretion, and of nutrition, including the circulatory, respiratory, and peristaltic movements of the heart, lungs, stomach, and intestines.

INDEX.

N.B.—The numbers refer to the respective SECTIONS.